Pocket Book of
Electrical Engineering
Formulas

Richard C. Dorf
Ronald J. Tallarida

CRC Press
Boca Raton   New York

Library of Congress Cataloging-in-Publication Data

Dorf, Richard C.
   Pocket book of electrical engineering formulas /Richard
C. Dorf and Ronald J. Tallarida
      p.   cm.
   Includes bibliographical references and index.
   ISBN 0-8493-4473-5
   1. Electric engineering—Mathematics—Handbooks,
manuals, etc.. 2. Mathematics—Formulae.   I. Tallarida,
Ronald J.   II. Title.
TK153.D68   1993
621.3′.0212—dc20   93-9634
                                                                              CIP

International Standard Book Number 0-8493-4473-5
Library of Congress Card Number 93-9634
Printed in the United States of America   4 5 6 7 8 9 0
Printed on acid-free paper

Preface

The purpose of this book is to serve the reference needs of electrical engineers. The material has been compiled so that it may serve the needs of students and professionals who wish to have a ready reference to formulas, equations, methods, concepts, and their mathematical formulation.

The contents and size make it especially convenient and portable. The widespread availability and low price of scientific calculators have greatly reduced the need for many numerical tables. Accordingly, this book contains the informaton required by electrical engineers. Sections 1 through 13 cover the key mathematical concepts and formulas used by most electrical engineers. Sections 14 through 31 cover the wide range of subjects normally included as the basics of electrical engineering.

The size of the book is comparable to that of many calculators and it is really very much a companion to the calculator and the computer as a source of information for writing one's own programs. To facilitate such use, the authors and the publisher have worked together to make the format attractive and clear.

Students and professionals alike will find this book a valuable supplement to standard textbooks, a source for review, and a handy reference for many years.

Ronald J. Tallarida
Philadelphia, PA

Richard C. Dorf
Davis, CA

About the Authors

Richard C. Dorf, professor of electrical and computer engineering at the University of California, Davis, teaches graduate and undergraduate courses in electrical engineering in the fields of circuits and control systems. He earned a Ph.D. in electrical engineering from the U.S. Naval Postgraduate School, an M.S. from the University of Colorado, and a B.S. from Clarkson University. Highly concerned with the discipline of electrical engineering and its wide value to social and economic needs, he has written and lectured internationally on the contributions and advances in electrical engineering.

Professor Dorf has extensive experience with education and industry and is professionally active in the fields of robotics, automation, electric circuits, and communications. He has served as a visiting professor at the University of Edinburgh, Scotland; the Massachusetts Institute of Technology; Stanford University; and the University of California, Berkeley.

A Fellow of the Institute of Electrical and Electronics Engineers, Dr. Dorf is widely known to the profession for his *Modern Control Systems*, 6th edition (Addison-Wesley, 1992), and *The International Encyclopedia of Robotics* (Wiley, 1988). Dr. Dorf is also the co-author of *Circuits, Devices and Systems* (with Ralph Smith), 5th edition (Wiley, 1992) and Editor-in-Chief of *The Electrical Engineering Handbook* (CRC Press, 1993).

Ronald J. Tallarida holds B.S. and M.S. degrees in physics/mathematics and a Ph.D. in pharmacology. His primary appointment is as Professor of Pharmacology at Temple University School of Medicine, Philadelphia; he also serves as Adjunct Professor of Biomedical Engineering (Mathematics) at Drexel University in Philadelphia.

He received the Lindback Award for Distinguished Teaching in 1964 while in the Drexel mathematics department. As an author and researcher, Dr. Tallarida has published over 150 works, including 7 books. He is currently the series editor for the Springer-Verlag Series in Pharmacologic Science.

Greek Letters

α	A	Alpha
β	B	Beta
γ	Γ	Gamma
δ	Δ	Delta
ϵ	E	Epsilon
ζ	Z	Zeta
η	H	Eta
θ	Θ	Theta
ι	I	Iota
κ	K	Kappa
λ	Λ	Lambda
μ	M	Mu
ν	N	Nu
ξ	Ξ	Xi
o	O	Omicron
π	Π	Pi
ρ	P	Rho
σ	Σ	Sigma
τ	T	Tau
υ	Υ	Upsilon
ϕ	Φ	Phi
χ	X	Chi
ψ	Ψ	Psi
ω	Ω	Omega

The Numbers π and e

π	=	3.14159	26535	89793
e	=	2.71828	18284	59045
$\log_{10}e$	=	0.43429	44819	03252
$\log_e 10$	=	2.30258	50929	94046

Prime Numbers

2	3	5	7	11	13	17	19	23	29
31	37	41	43	47	53	59	61	67	71
73	79	83	89	97	101	103	107	109	113
127	131	137	139	149	151	157	163	167	173
179	181	191	193	197	199	211	223	227	229
233	239	241	251	257	263	269	271	277	281
...				...					...

Important Numbers in Science (Physical Constants)

Avogadro constant (N_A)	6.02×10^{26} kmole^{-1}
Boltzmann constant (k)	1.38×10^{-23} J$\cdot$°K^{-1}
Electron charge (e)	1.602×10^{-19} C
Electron, charge/mass, (e/m_e)	1.760×10^{11} C$\cdot$kg^{-1}
Electron rest mass (m_e)	9.11×10^{-31} kg (0.511 MeV)
Faraday constant (F)	9.65×10^4 C$\cdot$mole^{-1}
Gas constant (R)	8.31×10^3 J$\cdot$°K$^{-1}\cdot$kmole^{-1}
Gas (ideal) normal volume (V_o)	22.4 m$^3\cdot$kmole^{-1}
Gravitational constant (G)	6.67×10^{-11} N$\cdot$m$^2\cdot$kg^{-2}
Hydrogen atom (rest mass) (m_H)	1.673×10^{-27} kg (938.8 MeV)

Neutron (rest mass)
  (m_n) 1.675×10^{-27} kg
                                (939.6 MeV)
Planck constant (h) 6.63×10^{-34} J·s
Proton (rest mass) (m_p) 1.673×10^{-27} kg
                                (938.3 MeV)
Speed of light (c) 3.00×10^8 m·s^{-1}

Contents

1 Elementary Algebra and Geometry

2 Determinants, Matrices, and Linear Systems of Equations

3 Trigonometry

4 Analytic Geometry

30 Digital Logic

31 Communication Systems

1 Elementary Algebra and Geometry

Algebra

1. Fundamental Properties (Real Numbers)

$a + b = b + a$	Commutative Law for Addition
$(a + b) + c = a + (b + c)$	Associative Law for Addition
$a + 0 = 0 + a$	Identity Law for Addition
$a + (-a) = (-a) + a = 0$	Inverse Law for Addition
$a(bc) = (ab)c$	Associative Law for Multiplication
$a\left(\dfrac{1}{a}\right) = \left(\dfrac{1}{a}\right)a = 1,\ a \neq 0$	Inverse Law for Multiplication
$(a)(1) = (1)(a) = a$	Identity Law for Multiplication
$ab = ba$	Commutative Law for Multiplication
$a(b + c) = ab + ac$	Distributive Law

DIVISION BY ZERO IS NOT DEFINED

2. Exponents

For integers m and n

$$a^n a^m = a^{n+m}$$

$$a^n / a^m = a^{n-m}$$

$$(a^n)^m = a^{nm}$$

$$(ab)^m = a^m b^m$$

$$(a/b)^m = a^m / b^m$$

3. Fractional Exponents

$$a^{p/q} = (a^{1/q})^p$$

where $a^{1/q}$ is the positive qth root of a if $a > 0$ and the negative qth root of a if a is negative and q is odd. Accordingly, the five rules of exponents given above (for integers) are also valid if m and n are fractions, provided a and b are positive.

4. Irrational Exponents

If an exponent is irrational, e.g., $\sqrt{2}$, the quantity, such as $a^{\sqrt{2}}$, is the limit of the sequence $a^{1.4}, a^{1.41}, a^{1.414}, \ldots$.

- ### Operations with Zero

$$0^m = 0; \qquad a^0 = 1$$

5. Logarithms

If x, y, and b are positive and $b \neq 1$

$$\log_b(xy) = \log_b x + \log_b y$$

$$\log_b(x/y) = \log_b x - \log_b y$$

$$\log_b x^p = p \log_b x$$

$$\log_b(1/x) = -\log_b x$$

$$\log_b b = 1$$

$$\log_b 1 = 0 \qquad Note: b^{\log_b x} = x.$$

- *Change of Base* $(a \neq 1)$

$$\log_b x = \log_a x \, \log_b a$$

6. Factorials

The factorial of a positive integer n is the product of all the positive integers less than or equal to the integer n and is denoted $n!$. Thus,

$$n! = 1 \cdot 2 \cdot 3 \cdot \ldots \cdot n.$$

Factorial 0 is defined: $0! = 1$.

- *Stirling's Approximation*

$$\lim_{n \to \infty} (n/e)^n \sqrt{2\pi n} = n!$$

(See also 9.2.)

7. Binomial Theorem

For positive integer n

$$(x+y)^n = x^n + nx^{n-1}y + \frac{n(n-1)}{2!}x^{n-2}y^2$$

$$+ \frac{n(n-1)(n-2)}{3!}x^{n-3}y^3 + \cdots$$

$$+ nxy^{n-1} + y^n.$$

8. Factors and Expansion

$$(a+b)^2 = a^2 + 2ab + b^2$$

$$(a-b)^2 = a^2 - 2ab + b^2$$

$$(a+b)^3 = a^3 + 3a^2b + 3ab^2 + b^3$$

$$(a-b)^3 = a^3 - 3a^2b + 3ab^2 - b^3$$

$$(a^2 - b^2) = (a-b)(a+b)$$

$$(a^3 - b^3) = (a-b)(a^2 + ab + b^2)$$

$$(a^3 + b^3) = (a+b)(a^2 - ab + b^2)$$

9. Progression

An *arithmetic progression* is a sequence in which the difference between any term and the preceding term is a constant (d):

$$a, a+d, a+2d, \ldots, a+(n-1)d.$$

If the last term is denoted l $[=a+(n-1)d]$, then the sum is

$$s = \frac{n}{2}(a+l)$$

A *geometric progression* is a sequence in which the ratio of any term to the preceding term is a constant r. Thus, for n terms

$$a, ar, ar^2, \ldots, ar^{n-1}$$

the sum is

$$S = \frac{a - ar^n}{1 - r}$$

10. Complex Numbers

A complex number is an ordered pair of real numbers (a, b).

Equality: $(a, b) = (c, d)$ if and only if $a = c$ and $b = d$
Addition: $(a, b) + (c, d) = (a+c, b+d)$
Multiplication: $(a, b)(c, d) = (ac - bd, ad + bc)$

The first element (a, b) is called the *real* part; the second the *imaginary* part. An alternate notation for (a, b) is $a + bi$, where $i^2 = (-1, 0)$, and $i = (0, 1)$ or $0 + 1i$ is written for this complex number as a convenience. With this understanding, i behaves as a number, i.e., $(2 - 3i)(4 + i) = 8 - 12i + 2i - 3i^2 = 11 - 10i$. The conjugate of $a + bi$ is $a - bi$ and the product of a complex number and its conjugate is $a^2 + b^2$. Thus, *quotients* are computed by multiplying numerator and denominator by the conjugate of the denominator, as

5

illustrated below:

$$\frac{2+3i}{4+2i} = \frac{(4-2i)(2+3i)}{(4-2i)(4+2i)} = \frac{14+8i}{20} = \frac{7+4i}{10}$$

11. Polar Form

The complex number $x+iy$ may be represented by a plane vector with components x and y

$$x+iy = r(\cos\theta + i\sin\theta)$$

(see Figure 1.1). Then, given two complex numbers $z_1 = r_1(\cos\theta_1 + i\sin\theta_1)$ and $z_2 = r_2(\cos\theta_2 + i\sin\theta_2)$, the product and quotient are

product: $z_1 z_2 = r_1 r_2 [\cos(\theta_1 + \theta_2) + i\sin(\theta_1 + \theta_2)]$

quotient: $z_1/z_2 = (r_1/r_2)[\cos(\theta_1 - \theta_2)$
$\qquad\qquad + i\sin(\theta_1 - \theta_2)]$

powers: $z^n = [r(\cos\theta + i\sin\theta)]^n$
$\qquad\qquad = r^n[\cos n\theta + i\sin n\theta]$

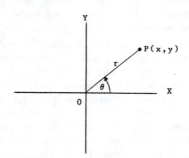

FIGURE 1.1. Polar form of complex number.

roots:
$$z^{1/n} = [r(\cos\theta + i\sin\theta)]^{1/n}$$
$$= r^{1/n}\left[\cos\frac{\theta + k\cdot 360}{n} + i\sin\frac{\theta + k\cdot 360}{n}\right],$$
$$k = 0, 1, 2, \ldots, n-1$$

12. Permutations

A permutation is an ordered arrangement (sequence) of all or part of a set of objects. The number of permutations of n objects taken r at a time is

$$p(n,r) = n(n-1)(n-2)\ldots(n-r+1)$$

$$= \frac{n!}{(n-r)!}$$

A permutation of positive integers is "even" or "odd" if the total number of inversions is an even integer or an odd integer, respectively. Inversions are counted rela - tive to each integer j in the permutation by counting the number of integers that follow j and are less than j. These are summed to give the total number of inversions. For example, the permutation 4132 has four inversions: three relative to 4 and one relative to 3. This permutation is therefore even.

13. Combinations

A combination is a selection of one or more objects from among a set of objects regardless of order. The number of combinations of n different objects taken r at a time is

$$C(n,r) = \frac{P(n,r)}{r!} = \frac{n!}{r!(n-r)!}$$

14. Algebraic Equations

- ### Quadratic

If $ax^2 + bx + c = 0$, and $a \neq 0$, then roots are

$$x = \frac{-b \pm \sqrt{b^2 - 4ac}}{2a}$$

- ### Cubic

To solve $x^3 + bx^2 + cx + d = 0$, let $x = y - b/3$. Then the *reduced cubic* is obtained:

$$y^3 + py + q = 0$$

where $p = c - (1/3)b^2$ and $q = d - (1/3)bc + (2/27)b^3$. Solutions of the original cubic are then in terms of the reduced cubic roots y_1, y_2, y_3:

$$x_1 = y_1 - (1/3)b \qquad x_2 = y_2 - (1/3)b$$

$$x_3 = y_3 - (1/3)b$$

The three roots of the reduced cubic are

$$y_1 = (A)^{1/3} + (B)^{1/3}$$

$$y_2 = W(A)^{1/3} + W^2(B)^{1/3}$$

$$y_3 = W^2(A)^{1/3} + W(B)^{1/3}$$

where

$$A = -\frac{1}{2}q + \sqrt{(1/27)p^3 + \frac{1}{4}q^2}\,,$$

$$B = -\frac{1}{2}q - \sqrt{(1/27)p^3 + \frac{1}{4}q^2}\,,$$

$$W = \frac{-1 + i\sqrt{3}}{2}\,,\quad W^2 = \frac{-1 - i\sqrt{3}}{2}\,.$$

When $(1/27)p^3 + (1/4)q^2$ is negative, A is complex; in this case A should be expressed in trigonometric form: $A = r(\cos\theta + i\sin\theta)$ where θ is a first or second quadrant angle, as q is negative or positive. The three roots of the reduced cubic are

$$y_1 = 2(r)^{1/3}\cos(\theta/3)$$

$$y_2 = 2(r)^{1/3}\cos\left(\frac{\theta}{3} + 120°\right)$$

$$y_3 = 2(r)^{1/3}\cos\left(\frac{\theta}{3} + 240°\right)$$

15. Geometry

The following is a collection of common geometric figures. Area (A), volume (V), and other measurable features are indicated.

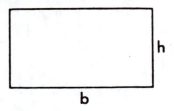

FIGURE 1.2. Rectangle. $A = bh$.

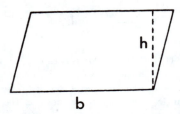

FIGURE 1.3. Parallelogram. $A = bh$.

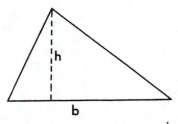

FIGURE 1.4. Triangle. $A = \frac{1}{2}bh$.

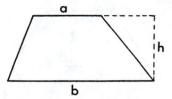

FIGURE 1.5. Trapezoid. $A = \frac{1}{2}(a+b)h$.

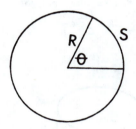

FIGURE 1.6. Circle. $A = \pi R^2$; circumference $= 2\pi R$; arc length $S = R\theta$ (θ in radians).

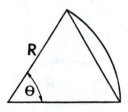

FIGURE 1.7. Sector of circle. $A_{\text{sector}} = \frac{1}{2}R^2\theta$; $A_{\text{segment}} = \frac{1}{2}R^2(\theta - \sin\theta)$.

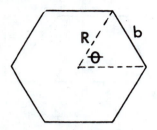

FIGURE 1.8. Regular polygon of n sides. $A = \frac{n}{4} b^2 \operatorname{ctn} \frac{\pi}{n}$; $R = \frac{b}{2} \csc \frac{\pi}{n}$.

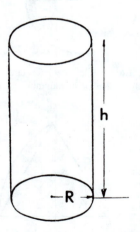

FIGURE 1.9. Right circular cylinder. $V = \pi R^2 h$; lateral surface area $= 2\pi Rh$.

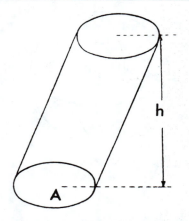

FIGURE 1.10. Cylinder (or prism) with parallel bases. $V = Ah$.

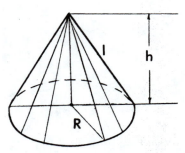

FIGURE 1.11. Right circular cone. $V = \frac{1}{3}\pi R^2 h$; lateral surface area $= \pi R l = \pi R \sqrt{R^2 + h^2}$.

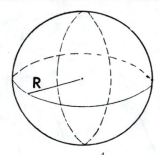

FIGURE 1.12. Sphere. $V = \dfrac{4}{3}\pi R^3$; surface area = $4\pi R^2$.

2

Determinants, Matrices, and Linear Systems of Equations

1. Determinants

Definition. The square array (matrix) A, with n rows and n columns, has associated with it the determinant

$$\det A = \begin{vmatrix} a_{11} & a_{12} & \cdots & a_{1n} \\ a_{21} & a_{22} & \cdots & a_{2n} \\ \cdots & \cdots & \cdots & \cdots \\ a_{n1} & a_{n2} & \cdots & a_{nn} \end{vmatrix},$$

a number equal to

$$\sum (\pm) a_{1i} a_{2j} a_{3k} \ldots a_{nl}$$

where $i, j, k, \ldots, l$ is a permutation of the n integers $1, 2, 3, \ldots, n$ in some order. The sign is plus if the permutation is *even* and is minus if the permutation is *odd* (see 1.12). The 2×2 determinant

$$\begin{vmatrix} a_{11} & a_{12} \\ a_{21} & a_{22} \end{vmatrix}$$

has the value $a_{11}a_{22} - a_{12}a_{21}$ since the permutation $(1, 2)$ is even and $(2, 1)$ is odd. For 3×3 determinants, permutations are as follows:

1,	2,	3	even
1,	3,	2	odd
2,	1,	3	odd
2,	3,	1	even
3,	1,	2	even
3,	2,	1	odd

Thus,

$$\begin{vmatrix} a_{11} & a_{12} & a_{13} \\ a_{21} & a_{22} & a_{23} \\ a_{31} & a_{32} & a_{33} \end{vmatrix} = \begin{cases} +a_{11} & \cdot & a_{22} & \cdot & a_{33} \\ -a_{11} & \cdot & a_{23} & \cdot & a_{32} \\ -a_{12} & \cdot & a_{21} & \cdot & a_{33} \\ +a_{12} & \cdot & a_{23} & \cdot & a_{31} \\ +a_{13} & \cdot & a_{21} & \cdot & a_{32} \\ -a_{13} & \cdot & a_{22} & \cdot & a_{31} \end{cases}$$

A determinant of order n is seen to be the sum of $n!$ signed products.

2. Evaluation by Cofactors

Each element a_{ij} has a determinant of order $(n-1)$ called a *minor* (M_{ij}) obtained by suppressing all elements in row i and column j. For example, the minor of element a_{22} in the 3×3 determinant above is

$$\begin{vmatrix} a_{11} & a_{13} \\ a_{31} & a_{33} \end{vmatrix}$$

The cofactor of element a_{ij}, denoted A_{ij}, is defined as $\pm M_{ij}$, where the sign is determined from i and j:

$$A_{ij} = (-1)^{i+j} M_{ij}.$$

16

The value of the $n \times n$ determinant equals the sum of products of elements of any row (or column) and their respective cofactors. Thus, for the 3×3 determinant

$$\det A = a_{11}A_{11} + a_{12}A_{12} + a_{13}A_{13} \text{ (first row)}$$

or

$$= a_{11}A_{11} + a_{21}A_{21} + a_{31}A_{31} \text{ (first column)}$$

etc.

3. Properties of Determinants

a. If the corresponding columns and rows of A are interchanged, det A is unchanged.

b. If any two rows (or columns) are interchanged, the sign of det A changes.

c. If any two rows (or columns) are identical, det $A = 0$.

d. If A is triangular (all elements above the main diagonal equal to zero), $A = a_{11} \cdot a_{22} \cdot \ldots \cdot a_{nn}$:

$$
\begin{vmatrix}
a_{11} & 0 & 0 & \cdots & 0 \\
a_{21} & a_{22} & 0 & \cdots & 0 \\
\cdots & \cdots & \cdots & \cdots & \cdots \\
a_{n1} & a_{n2} & a_{n3} & \cdots & a_{nn}
\end{vmatrix}
$$

e. If to each element of a row or column there is added C times the corresponding element in another row (or column), the value of the determinant is unchanged.

4. Matrices

Definition. A matrix is a rectangular array of numbers and is represented by a symbol A or $[a_{ij}]$:

$$A = \begin{bmatrix} a_{11} & a_{12} & \cdots & a_{1n} \\ a_{21} & a_{22} & \cdots & a_{2n} \\ \cdots & \cdots & \cdots & \cdots \\ a_{m1} & a_{m2} & \cdots & a_{mn} \end{bmatrix} = [a_{ij}]$$

The numbers a_{ij} are termed *elements* of the matrix; subscripts i and j identify the element as the number in row i and column j. The order of the matrix is $m \times n$ ("m by n"). When $m = n$, the matrix is square and is said to be of order n. For a square matrix of order n the elements $a_{11}, a_{22}, \ldots, a_{nn}$ constitute the main diagonal.

5. Operations

Addition. Matrices A and B of the same order may be added by adding corresponding elements, i.e., $A + B = [(a_{ij} + b_{ij})]$.

Scalar multiplication. If $A = [a_{ij}]$ and c is a constant (scalar), then $cA = [ca_{ij}]$, that is, every element of A is multiplied by c. In particular, $(-1)A = -A = [-a_{ij}]$ and $A + (-A) = 0$, a matrix with all elements equal to zero.

Multiplication of matrices. Matrices A and B may be multiplied only when they are conformable, which means that the number of columns of A equals the number of rows of B. Thus, if A is $m \times k$ and B is $k \times n$, then the product $C = AB$ exists as an $m \times n$ matrix with elements c_{ij} equal to the sum of products of elements in row

i of A and corresponding elements of column j of B:

$$c_{ij} = \sum_{l=1}^{k} a_{il}b_{lj}$$

For example, if

$$\begin{bmatrix} a_{11} & a_{12} & \cdots & a_{1k} \\ a_{21} & a_{22} & \cdots & a_{2k} \\ \cdots & \cdots & \cdots & \cdots \\ a_{m1} & \cdots & \cdots & a_{mk} \end{bmatrix} \cdot \begin{bmatrix} b_{11} & b_{12} & \cdots & b_{1n} \\ b_{21} & b_{22} & \cdots & b_{2n} \\ \cdots & \cdots & \cdots & \cdots \\ b_{k1} & b_{k2} & \cdots & b_{kn} \end{bmatrix}$$

$$= \begin{bmatrix} c_{11} & c_{12} & \cdots & c_{1n} \\ c_{21} & c_{22} & \cdots & c_{2n} \\ \cdots & \cdots & \cdots & \\ c_{m1} & c_{m2} & \cdots & c_{mn} \end{bmatrix}$$

then element c_{21} is the sum of products $a_{21}b_{11} + a_{22}b_{21} + \ldots + a_{2k}b_{k1}$.

6. Properties

$$A + B = B + A$$
$$A + (B + C) = (A + B) + C$$
$$(c_1 + c_2)A = c_1 A + c_2 A$$
$$c(A + B) = cA + cB$$
$$c_1(c_2 A) = (c_1 c_2)A$$
$$(AB)(C) = A(BC)$$
$$(A + B)(C) = AC + BC$$
$$AB \neq BA \text{ (in general)}$$

19

7. Transpose

If A is an $n \times m$ matrix, the matrix of order $m \times n$ obtained by interchanging the rows and columns of A is called the *transpose* and is denoted A^T. The following are properties of A, B, and their respective transposes:

$$\left(A^T \right)^T = A$$

$$\left(A + B \right)^T = A^T + B^T$$

$$\left(cA \right)^T = cA^T$$

$$\left(AB \right)^T = B^T A^T$$

A *symmetric* matrix is a square matrix A with the property $A = A^T$.

8. Identity Matrix

A square matrix in which each element of the main diagonal is the same constant a and all other elements zero is called a *scalar* matrix.

$$\begin{bmatrix} a & 0 & 0 & \cdots & 0 \\ 0 & a & 0 & \cdots & 0 \\ 0 & 0 & a & \cdots & 0 \\ \cdots & \cdots & \cdots & \cdots & \\ 0 & 0 & 0 & \cdots & a \end{bmatrix}$$

When a scalar matrix multiplies a conformable second matrix A, the product is aA; that is, the same as multiplying A by a scalar a. A scalar matrix with diagonal elements 1 is called the *identity*, or *unit* matrix and is denoted I. Thus, for any nth order matrix A, the identity matrix of order n has the property

$$AI = IA = A$$

9. Adjoint

If A is an n-order square matrix and A_{ij} the cofactor of element a_{ij}, the transpose of $[A_{ij}]$ is called the *adjoint* of A:

$$\text{adj } A = [A_{ij}]^T$$

10. Inverse Matrix

Given a square matrix A of order n, if there exists a matrix B such that $AB = BA = I$, then B is called the *inverse* of A. The inverse is denoted A^{-1}. A necessary and sufficient condition that the square matrix A have an inverse is $\det A \neq 0$. Such a matrix is called *nonsingular*; its inverse is unique and it is given by

$$A^{-1} = \frac{\text{adj } A}{\det A}$$

Thus, to form the inverse of the nonsingular matrix A, form the adjoint of A and divide each element of the adjoint by $\det A$. For example,

$$\begin{bmatrix} 1 & 0 & 2 \\ 3 & -1 & 1 \\ 4 & 5 & 6 \end{bmatrix} \text{ has matrix of cofactors}$$

$$\begin{bmatrix} -11 & -14 & 19 \\ 10 & -2 & -5 \\ 2 & 5 & -1 \end{bmatrix},$$

$$\text{adjoint} = \begin{bmatrix} -11 & 10 & 2 \\ -14 & -2 & 5 \\ 19 & -5 & -1 \end{bmatrix} \text{ and determinant 27.}$$

Therefore,

$$A^{-1} = \begin{bmatrix} \dfrac{-11}{27} & \dfrac{10}{27} & \dfrac{2}{27} \\[2mm] \dfrac{-14}{27} & \dfrac{-2}{27} & \dfrac{5}{27} \\[2mm] \dfrac{19}{27} & \dfrac{-5}{27} & \dfrac{-1}{27} \end{bmatrix}.$$

11. Systems of Linear Equations

Given the system

$$
\begin{array}{ccccccccc}
a_{11}x_1 & + & a_{12}x_2 & + \cdots + & a_{1n}x_n & = & b_1 \\
a_{21}x_1 & + & a_{22}x_2 & + \cdots + & a_{2n}x_n & = & b_2 \\
\vdots & & \vdots & \vdots & \vdots & & \vdots \\
a_{n1}x_1 & + & a_{n2}x_2 & + \cdots + & a_{nn}x_n & = & b_n
\end{array}
$$

a unique solution exists if $\det A \neq 0$, where A is the $n \times n$ matrix of coefficients $[a_{ij}]$.

- *Solution by Determinants (Cramer's Rule)*

$$
x_1 = \begin{vmatrix} b_1 & a_{12} & \cdots & a_{1n} \\ b_2 & a_{22} & & \\ \vdots & \vdots & & \vdots \\ b_n & a_{n2} & & a_{nn} \end{vmatrix} \div \det A
$$

$$x_2 = \begin{vmatrix} a_{11} & b_1 & a_{13} & \cdots & a_{1n} \\ a_{21} & b_2 & \cdots & & \cdots \\ \vdots & \vdots & & & \\ a_{n1} & b_n & a_{n3} & & a_{nn} \end{vmatrix} \div \det A$$

$$\vdots$$

$$x_k = \frac{\det A_k}{\det A},$$

where A_k is the matrix obtained from A by replacing the kth column of A by the column of b's.

12. Matrix Solution

The linear system may be written in matrix form $AX = B$ where A is the matrix of coefficients $[a_{ij}]$ and X and B are

$$X = \begin{bmatrix} x_1 \\ x_2 \\ \vdots \\ x_n \end{bmatrix} \qquad B = \begin{bmatrix} b_1 \\ b_2 \\ \vdots \\ b_n \end{bmatrix}$$

If a unique solution exists, $\det A \neq 0$; hence A^{-1} exists and

$$X = A^{-1}B.$$

23

3 Trigonometry

1. Triangles

In any triangle (in a plane) with sides a, b, and c and corresponding opposite angles A, B, C,

$$\frac{a}{\sin A} = \frac{b}{\sin B} = \frac{c}{\sin C}$$ Law of Sines

$$a^2 = b^2 + c^2 - 2cb \cos A$$ Law of Cosines

$$\frac{a+b}{a-b} = \frac{\tan \frac{1}{2}(A+B)}{\tan \frac{1}{2}(A-B)}$$ Law of Tangents

$$\sin \frac{1}{2} A = \sqrt{\frac{(s-b)(s-c)}{bc}}, \quad \text{where } s = \frac{1}{2}(a+b+c).$$

$$\cos \frac{1}{2} A = \sqrt{\frac{s(s-a)}{bc}}.$$

$$\tan \frac{1}{2} A = \sqrt{\frac{(s-b)(s-c)}{s(s-a)}}.$$

$$\text{Area} = \frac{1}{2} bc \sin A$$
$$= \sqrt{s(s-a)(s-b)(s-c)}.$$

If the vertices have coordinates $(x_1, y_1), (x_2, y_2),$ (x_3, y_3), the area is the *absolute value* of the expression

$$\frac{1}{2} \begin{vmatrix} x_1 & y_1 & 1 \\ x_2 & y_2 & 1 \\ x_3 & y_3 & 1 \end{vmatrix}$$

2. Trigonometric Functions of an Angle

With reference to Figure 3.1, $P(x, y)$ is a point in either one of the four quadrants and A is an angle whose initial side is coincident with the positive x-axis and whose terminal side contains the point $P(x, y)$. The distance from the origin $P(x, y)$ is denoted by r and is positive. The trigonometric functions of the

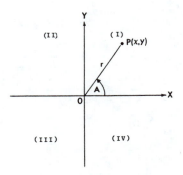

FIGURE 3.1. The trigonometric point. Angle A is taken to be positive when the rotation is counterclockwise and negative when the rotation is clockwise. The plane is divided into quadrants as shown.

angle A are defined as:

$$\sin A = \text{sine } A = y/r$$
$$\cos A = \text{cosine } A = x/r$$
$$\tan A = \text{tangent } A = y/x$$
$$\text{ctn } A = \text{cotangent } A = x/y$$
$$\sec A = \text{secant } A = r/x$$
$$\csc A = \text{cosecant } A = r/y$$

Angles are measured in degrees or radians; $180° = \pi$ radians; 1 radian $= 180°/\pi$ degrees.

The trigonometric functions of 0°, 30°, 45°, and integer multiples of these are directly computed.

	0°	30°	45°	60°	90°	120°	135°	150°	180°
sin	0	$\frac{1}{2}$	$\frac{\sqrt{2}}{2}$	$\frac{\sqrt{3}}{2}$	1	$\frac{\sqrt{3}}{2}$	$\frac{\sqrt{2}}{2}$	$\frac{1}{2}$	0
cos	1	$\frac{\sqrt{3}}{2}$	$\frac{\sqrt{2}}{2}$	$\frac{1}{2}$	0	$-\frac{1}{2}$	$-\frac{\sqrt{2}}{2}$	$-\frac{\sqrt{3}}{2}$	-1
tan	0	$\frac{\sqrt{3}}{3}$	1	$\sqrt{3}$	∞	$-\sqrt{3}$	-1	$-\frac{\sqrt{3}}{3}$	0
ctn	∞	$\sqrt{3}$	1	$\frac{\sqrt{3}}{3}$	0	$-\frac{\sqrt{3}}{3}$	-1	$-\sqrt{3}$	∞
sec	1	$\frac{2\sqrt{3}}{3}$	$\sqrt{2}$	2	∞	-2	$-\sqrt{2}$	$-\frac{2\sqrt{3}}{3}$	-1
csc	∞	2	$\sqrt{2}$	$\frac{2\sqrt{3}}{3}$	1	$\frac{2\sqrt{3}}{3}$	$\sqrt{2}$	2	∞

3. Trigonometric Identities

$$\sin A = \frac{1}{\csc A}$$

$$\cos A = \frac{1}{\sec A}$$

$$\tan A = \frac{1}{\operatorname{ctn} A} = \frac{\sin A}{\cos A}$$

$$\csc A = \frac{1}{\sin A}$$

$$\sec A = \frac{1}{\cos A}$$

$$\operatorname{ctn} A = \frac{1}{\tan A} = \frac{\cos A}{\sin A}$$

$$\sin^2 A + \cos^2 A = 1$$

$$1 + \tan^2 A = \sec^2 A$$

$$1 + \operatorname{ctn}^2 A = \csc^2 A$$

$$\sin(A \pm B) = \sin A \cos B \pm \cos A \sin B$$

$$\cos(A \pm B) = \cos A \cos B \mp \sin A \sin B$$

$$\tan(A \pm B) = \frac{\tan A \pm \tan B}{1 \mp \tan A \tan B}$$

$$\sin 2A = 2 \sin A \cos A$$

$$\sin 3A = 3 \sin A - 4 \sin^3 A$$

$$\sin nA = 2\sin(n-1)A \cos A - \sin(n-2)A$$

$$\cos 2A = 2\cos^2 A - 1 = 1 - 2\sin^2 A$$

$$\cos 3A = 4\cos^3 A - 3\cos A$$

$$\cos nA = 2\cos(n-1)A \cos A - \cos(n-2)A$$

$$\sin A + \sin B = 2\sin\frac{1}{2}(A+B)\cos\frac{1}{2}(A-B)$$

$$\sin A - \sin B = 2\cos\frac{1}{2}(A+B)\sin\frac{1}{2}(A-B)$$

$$\cos A + \cos B = 2\cos\frac{1}{2}(A+B)\cos\frac{1}{2}(A-B)$$

$$\cos A - \cos B = -2\sin\frac{1}{2}(A+B)\sin\frac{1}{2}(A-B)$$

$$\tan A \pm \tan B = \frac{\sin(A \pm B)}{\cos A \cos B}$$

$$\operatorname{ctn} A \pm \operatorname{ctn} B = \pm\frac{\sin(A \pm B)}{\sin A \sin B}$$

$$\sin A \sin B = \frac{1}{2}\cos(A-B) - \frac{1}{2}\cos(A+B)$$

$$\cos A \cos B = \frac{1}{2}\cos(A-B) + \frac{1}{2}\cos(A+B)$$

$$\sin A \cos B = \frac{1}{2}\sin(A+B) + \frac{1}{2}\sin(A-B)$$

$$\sin\frac{A}{2} = \pm\sqrt{\frac{1-\cos A}{2}}$$

$$\cos\frac{A}{2} = \pm\sqrt{\frac{1 + \cos A}{2}}$$

$$\tan\frac{A}{2} = \frac{1 - \cos A}{\sin A} = \frac{\sin A}{1 + \cos A} = \pm\sqrt{\frac{1 - \cos A}{1 + \cos A}}$$

$$\sin^2 A = \frac{1}{2}(1 - \cos 2A)$$

$$\cos^2 A = \frac{1}{2}(1 + \cos 2A)$$

$$\sin^3 A = \frac{1}{4}(3\sin A - \sin 3A)$$

$$\cos^3 A = \frac{1}{4}(\cos 3A + 3\cos A)$$

$$\sin ix = \frac{1}{2}i(e^x - e^{-x}) = i\sinh x$$

$$\cos ix = \frac{1}{2}(e^x + e^{-x}) = \cosh x$$

$$\tan ix = \frac{i(e^x - e^{-x})}{e^x + e^{-x}} = i\tanh x$$

$$e^{x+iy} = e^x(\cos y + i\sin y)$$

$$(\cos x \pm i\sin x)^n = \cos nx \pm i\sin nx$$

4. Inverse Trigonometric Functions

The inverse trigonometric functions are multiple valued, and this should be taken into account in the use of the following formulas.

$$\sin^{-1} x = \cos^{-1}\sqrt{1-x^2}$$

$$= \tan^{-1}\frac{x}{\sqrt{1-x^2}} = \operatorname{ctn}^{-1}\frac{\sqrt{1-x^2}}{x}$$

$$= \sec^{-1}\frac{1}{\sqrt{1-x^2}} = \csc^{-1}\frac{1}{x}$$

$$= -\sin^{-1}(-x)$$

$$\cos^{-1} x = \sin^{-1}\sqrt{1-x^2}$$

$$\cdot = \tan^{-1}\frac{\sqrt{1-x^2}}{x} = \operatorname{ctn}^{-1}\frac{x}{\sqrt{1-x^2}}$$

$$= \sec^{-1}\frac{1}{x} = \csc^{-1}\frac{1}{\sqrt{1-x^2}}$$

$$= \pi - \cos^{-1}(-x)$$

$$\tan^{-1} x = \operatorname{ctn}^{-1}\frac{1}{x}$$

$$= \sin^{-1}\frac{x}{\sqrt{1+x^2}} = \cos^{-1}\frac{1}{\sqrt{1+x^2}}$$

$$= \sec^{-1}\sqrt{1+x^2} = \csc^{-1}\frac{\sqrt{1+x^2}}{x}$$

$$= -\tan^{-1}(-x)$$

4 Analytic Geometry

1. Rectangular Coordinates

The points in a plane may be placed in one-to-one correspondence with pairs of real numbers. A common method is to use perpendicular lines that are horizontal and vertical and intersect at a point called the *origin*. These two lines constitute the coordinate axes; the horizontal line is the x-axis and the vertical line is the y-axis. The positive direction of the x-axis is to the right whereas the positive direction of the y-axis is up. If P is a point in the plane one may draw lines through it that are perpendicular to the x- and y-axes (such as the broken lines of Figure 4.1). The lines intersect the x-axis at a point with coordinate x_1 and the y-axis at a

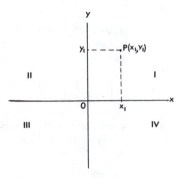

FIGURE 4.1. Rectangular coordinates.

31

point with coordinate y_1. We call x_1 the x-coordinate or *abscissa* and y_1 is termed the y-coordinate or *ordinate* of the point P. Thus, point P is associated with the pair of real numbers (x_1, y_1) and is denoted $P(x_1, y_1)$. The coordinate axes divide the plane into quadrants I, II, III, and IV.

2. Distance between Two Points; Slope

The distance d between the two points $P_1(x_1, y_1)$ and $P_2(x_2, y_2)$ is

$$d = \sqrt{(x_2 - x_1)^2 + (y_2 - y_1)^2}$$

In the special case when P_1 and P_2 are both on one of the coordinate axes, for instance, the x-axis,

$$d = \sqrt{(x_2 - x_1)^2} = |x_2 - x_1|,$$

or on the y-axis,

$$d = \sqrt{(y_2 - y_1)^2} = |y_2 - y_1|.$$

The midpoint of the line segment $P_1 P_2$ is

$$\left(\frac{x_1 + x_2}{2}, \frac{y_1 + y_2}{2} \right).$$

The slope of the line segment $P_1 P_2$, provided it is not vertical, is denoted by m and is given by

$$m = \frac{y_2 - y_1}{x_2 - x_1}.$$

The slope is related to the angle of inclination α (Figure 4.2) by

$$m = \tan \alpha$$

Two lines (or line segments) with slopes m_1 and m_2 are perpendicular if

$$m_1 = -1/m_2$$

and are parallel if $m_1 = m_2$.

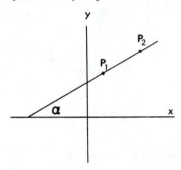

FIGURE 4.2. The angle of inclination is the smallest angle measured counterclockwise from the positive x-axis to the line that contains $P_1 P_2$.

3. Equations of Straight Lines

A *vertical* line has an equation of the form

$$x = c$$

where $(c, 0)$ is its intersection with the x-axis. A line of slope m through point (x_1, y_1) is given by

$$y - y_1 = m(x - x_1)$$

33

Thus, a *horizontal line* (slope $= 0$) through point (x_1, y_1) is given by

$$y = y_1.$$

A nonvertical line through the two points $P_1(x_1, y_1)$ and $P_2(x_2, y_2)$ is given by either

$$y - y_1 = \left(\frac{y_2 - y_1}{x_2 - x_1} \right)(x - x_1)$$

or

$$y - y_2 = \left(\frac{y_2 - y_1}{x_2 - x_1} \right)(x - x_2).$$

A line with x-intercept a and y-intercept b is given by

$$\frac{x}{a} + \frac{y}{b} = 1 \qquad (a \neq 0, \, b \neq 0).$$

The *general equation* of a line is

$$Ax + By + C = 0$$

The *normal form* of the straight line equation is

$$x \cos \theta + y \sin \theta = p$$

where p is the distance along the normal from the origin and θ is the angle that the normal makes with the x-axis (Figure 4.3).

The general equation of the line $Ax + By + C = 0$ may be written in normal form by dividing by $\pm \sqrt{A^2 + B^2}$, where the plus sign is used when C is negative and the

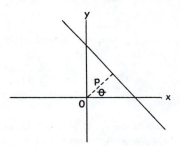

FIGURE 4.3. Construction for normal form of straight line equation.

minus sign is used when C is positive:

$$\frac{Ax + By + C}{\pm \sqrt{A^2 + B^2}} = 0,$$

so that

$$\cos \theta = \frac{A}{\pm \sqrt{A^2 + B^2}}, \qquad \sin \theta = \frac{B}{\pm \sqrt{A^2 + B^2}}$$

and

$$p = \frac{|C|}{\sqrt{A^2 + B^2}}.$$

4. Distance from a Point to a Line

The perpendicular distance from a point $P(x_1, y_1)$ to the line $Ax + By + C = 0$ is given by d

$$d = \frac{Ax_1 + By_1 + C}{\pm \sqrt{A^2 + B^2}}.$$

5. Circle

The general equation of a circle of radius r and center at $P(x_1, y_1)$ is

$$(x - x_1)^2 + (y - y_1)^2 = r^2.$$

6. Parabola

A parabola is the set of all points (x, y) in the plane that are equidistant from a given line called the *directrix* and a given point called the *focus*. The parabola is symmetric about a line that contains the focus and is perpendicular to the directrix. The line of symmetry intersects the parabola at its *vertex* (Figure 4.4). The eccentricity $e = 1$.

The distance between the focus and the vertex, or vertex and directrix, is denoted by p (> 0) and leads to one of the following equations of a parabola with vertex at the origin (Figures 4.5 and 4.6):

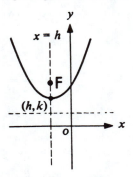

FIGURE 4.4. Parabola with vertex at (h, k). F identifies the focus.

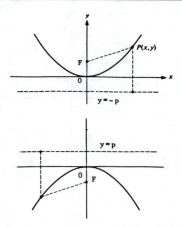

FIGURE 4.5. Parabolas with y-axis as the axis of symmetry and vertex at the origin. (Upper) $y = \dfrac{x^2}{4p}$; (lower) $y = -\dfrac{x^2}{4p}$.

$$y = \frac{x^2}{4p} \qquad \text{(opens upward)}$$

$$y = -\frac{x^2}{4p} \qquad \text{(opens downward)}$$

$$x = \frac{y^2}{4p} \qquad \text{(opens to right)}$$

$$x = -\frac{y^2}{4p} \qquad \text{(opens to left)}$$

37

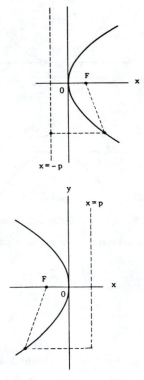

FIGURE 4.6. Parabolas with x-axis as the axis of symmetry and vertex at the origin. (Upper) $x = \dfrac{y^2}{4p}$; (lower) $x = -\dfrac{y^2}{4p}$.

38

For each of the four orientations shown in Figures 4.5 and 4.6, the corresponding parabola with vertex (h, k) is obtained by replacing x by $x - h$ and y by $y - k$. Thus, the parabola in Figure 4.7 has the equation

$$x - h = - \frac{(y - k)^2}{4p}.$$

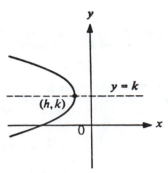

FIGURE 4.7. Parabola with vertex at (h, k) and axis parallel to the x-axis.

7. Ellipse

An ellipse is the set of all points in the plane such that the sum of their distances from two fixed points, called *foci*, is a given constant $2a$. The distance between the foci is denoted $2c$; the length of the major axis is $2a$, whereas the length of the minor axis is $2b$ (Figure 4.8) and

$$a = \sqrt{b^2 + c^2}.$$

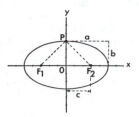

FIGURE 4.8. Ellipse; since point P is equidistant from foci F_1 and F_2 the segments F_1P and $F_2P = a$; hence $a = \sqrt{b^2 + c^2}$.

The eccentricity of an ellipse, e, is < 1. An ellipse with center at point (h, k) and major axis *parallel to the x-axis* (Figure 4.9) is given by the equation

$$\frac{(x-h)^2}{a^2} + \frac{(y-k)^2}{b^2} = 1.$$

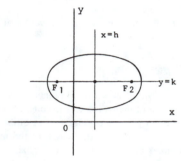

FIGURE 4.9. Ellipse with major axis parallel to the x-axis. F_1 and F_2 are the foci, each a distance c from center (h, k).

40

An ellipse with center at (h, k) and major axis parallel to the y-axis is given by the equation (Figure 4.10)

$$\frac{(y-k)^2}{a^2} + \frac{(x-h)^2}{b^2} = 1.$$

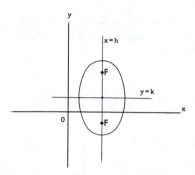

FIGURE 4.10. Ellipse with major axis parallel to the y-axis. Each focus is a distance c from center (h, k).

8. Hyperbola $(e > 1)$

A hyperbola is the set of all points in the plane such that the difference of its distances from two fixed points (foci) is a given positive constant denoted $2a$. The distance between the two foci is $2c$ and that between the two vertices is $2a$. The quantity b is defined by the equation

$$b = \sqrt{c^2 - a^2}$$

and is illustrated in Figure 4.11, which shows the construction of a hyperbola given by the equation

$$\frac{x^2}{a^2} - \frac{y^2}{b^2} = 1.$$

When the focal axis is parallel to the y-axis the equation of the hyperbola with center (h, k) (Figures 4.12
and 4.13) is

$$\frac{(y-k)^2}{a^2} - \frac{(x-h)^2}{b^2} = 1.$$

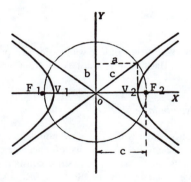

FIGURE 4.11. Hyperbola; V_1, V_2 = vertices; F_1, F_2 =
foci. A circle at center O with radius c contains the
vertices and illustrates the relation among a, b, and c.
Asymptotes have slopes b/a and $-b/a$ for the orien-
tation shown.

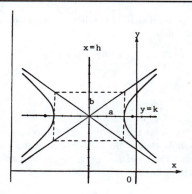

FIGURE 4.12. Hyperbola with center at (h, k):
$$\frac{(x-h)^2}{a^2} - \frac{(y-k)^2}{b^2} = 1; \text{ slopes of asymptotes } \pm b/a.$$

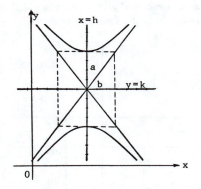

FIGURE 4.13. Hyperbola with center at (h, k):
$$\frac{(y-k)^2}{a^2} - \frac{(x-h)^2}{b^2} = 1; \text{ slopes of asymptotes } \pm a/b.$$

If the focal axis is parallel to the x-axis and center (h, k), then

$$\frac{(x-h)^2}{a^2} - \frac{(y-k)^2}{b^2} = 1$$

9. Change of Axes

A change in the position of the coordinate axes will generally change the coordinates of the points in the plane. The equation of a particular curve will also generally change.

- *Translation*

When the new axes remain parallel to the original, the transformation is called a *translation* (Figure 4.14). The new axes, denoted x' and y', have origin $0'$ at (h, k) with reference to the x and y axes.

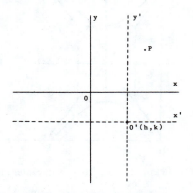

FIGURE 4.14. Translation of axes.

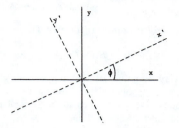

FIGURE 4.15. Rotation of axes.

A point P with coordinates (x, y) with respect to the original has coordinates (x', y') with respect to the new axes. These are related by

$$x = x' + h$$
$$y = y' + k$$

For example, the ellipse of Figure 4.10 has the following simpler equation with respect to axes x' and y' with the center at (h, k):

$$\frac{y'^2}{a^2} + \frac{x'^2}{b^2} = 1.$$

- *Rotation*

When the new axes are drawn through the same origin, remaining mutually perpendicular, but tilted with respect to the original, the transformation is one of rotation. For angle of rotation ϕ (Figure 4.15), the coordinates (x, y) and (x', y') of a point P are related by

$$x = x' \cos \phi - y' \sin \phi$$

$$y = x' \sin \phi + y' \cos \phi$$

10. General Equation of Degree Two

$$Ax^2 + Bxy + Cy^2 + Dx + Ey + F = 0$$

Every equation of the above form defines a conic section or one of the limiting forms of a conic. By rotating the axes through a particular angle ϕ, the xy-term vanishes, yielding

$$A'x'^2 + C'y'^2 + D'x' + E'y' + F' = 0$$

with respect to the axes x' and y'. The required angle ϕ (see Figure 4.15) is calculated from

$$\tan 2\phi = \frac{B}{A - C}, \qquad (\phi < 90°).$$

11. Polar Coordinates (Figure 4.16)

The fixed point O is the origin or *pole* and a line OA drawn through it is the polar axis. A point P in the plane is determined from its distance r, measured from

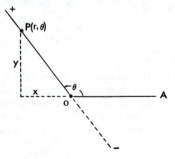

FIGURE 4.16. Polar coordinates.

46

O, and the angle θ between OP and OA. Distances measured on the terminal line of θ from the pole are positive, whereas those measured in the opposite direction are negative.

Rectangular coordinates (x, y) and polar coordinates (r, θ) are related according to

$$x = r \cos \theta, \qquad y = r \sin \theta$$

$$r^2 = x^2 + y^2, \qquad \tan \theta = y/x.$$

Several well-known polar curves are shown in Figures 4.17 to 4.21.

The polar equation of a conic section with focus at the pole and distance $2p$ from directrix to focus is either

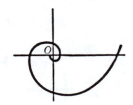

FIGURE 4.17. Polar curve $r = e^{a\theta}$.

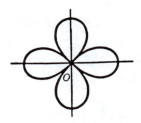

FIGURE 4.18. Polar curve $r = a \cos 2\theta$.

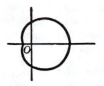

FIGURE 4.19. Polar curve $r = 2a \cos\theta + b$.

FIGURE 4.20. Polar curve $r = a \sin 3\theta$.

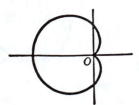

FIGURE 4.21. Polar curve $r = a(1 - \cos\theta)$.

$$r = \frac{2ep}{1 - e\cos\theta} \qquad \text{(directrix to left of pole)}$$

or

$$r = \frac{2ep}{1 + e\cos\theta} \qquad \text{(directrix to right of pole)}$$

The corresponding equations for the directrix below or above the pole are as above, except that sin θ appears instead of cos θ.

12. *Curves and Equations*

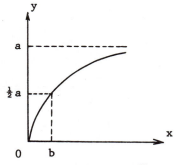

FIGURE 4.22. $y = \dfrac{ax}{x+b}$.

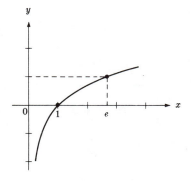

FIGURE 4.23. $y = \log x$.

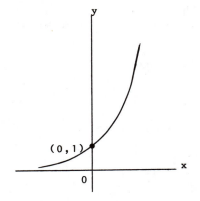

FIGURE 4.24. $y = e^x$.

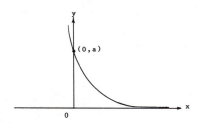

FIGURE 4.25. $y = ae^{-x}$.

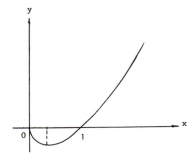

FIGURE 4.26. $y = x \log x$.

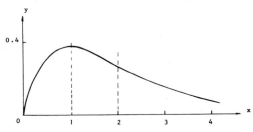

FIGURE 4.27. $y = xe^{-x}$.

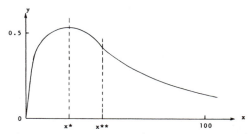

FIGURE 4.28. $y = e^{-ax} - e^{-bx}$, $0 < a < b$ (drawn for $a = 0.02$, $b = 0.1$, and showing maximum and inflection).

51

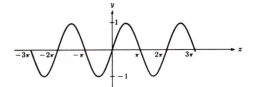

FIGURE 4.29. $y = \sin x$.

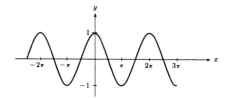

FIGURE 4.30. $y = \cos x$.

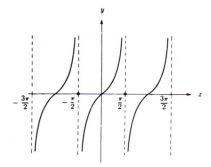

FIGURE 4.31. $y = \tan x$.

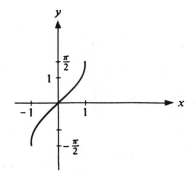

FIGURE 4.32. $y = \arcsin x$.

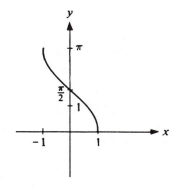

FIGURE 4.33. $y = \arccos x$.

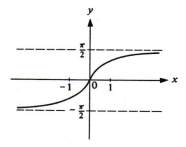

FIGURE 4.34. $y = \arctan x$.

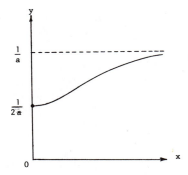

FIGURE 4.35. $y = e^{bx}/a(1+e^{bx})$, $x \geq 0$
(logistic equation).

54

5 Series

1. Bernoulli and Euler Numbers

A set of numbers, $B_1, B_3, \ldots, B_{2n-1}$ (Bernoulli numbers) and $B_2, B_4, \ldots, B_{2n}$ (Euler numbers) appear in the series expansions of many functions. A partial listing follows; these are computed from the following equations:

$$B_{2n} - \frac{2n(2n-1)}{2!} B_{2n-2}$$

$$+ \frac{2n(2n-1)(2n-2)(2n-3)}{4!} B_{2n-4} - \cdots$$

$$+ (-1)^n = 0,$$

and

$$\frac{2^{2n}(2^{2n}-1)}{2n} B_{2n-1} = (2n-1)B_{2n-2}$$

$$- \frac{(2n-1)(2n-2)(2n-3)}{3!} B_{2n-4} + \cdots + (-1)^{n-1}.$$

$$
\begin{array}{ll}
B_1 = 1/6 & B_2 = 1 \\
B_3 = 1/30 & B_4 = 5 \\
B_5 = 1/42 & B_6 = 61 \\
B_7 = 1/30 & B_8 = 1385 \\
B_9 = 5/66 & B_{10} = 50521
\end{array}
$$

$$B_{11} = 691/2730 \qquad B_{12} = 2702765$$
$$B_{13} = 7/6 \qquad B_{14} = 199360981$$
$$\vdots \qquad\qquad \vdots$$

2. Series of Functions

In the following, the interval of convergence is indicated, otherwise it is all x. Logarithms are to the base e. Bernoulli and Euler numbers (B_{2n-1} and B_{2n}) appear in certain expressions.

$$(a+x)^n = a^n + na^{n-1}x + \frac{n(n-1)}{2!}a^{n-2}x^2$$

$$+ \frac{n(n-1)(n-2)}{3!}a^{n-3}x^3 + \dots$$

$$+ \frac{n!}{(n-j)!j!}a^{n-j}x^j + \dots \quad [x^2 < a^2]$$

$$(a-bx)^{-1} = \frac{1}{a}\left[1 + \frac{bx}{a} + \frac{b^2x^2}{a^2} + \frac{b^3x^3}{a^3} + \dots\right]$$

$$[b^2x^2 < a^2]$$

$$(1 \pm x)^n = 1 \pm nx + \frac{n(n-1)}{2!}x^2$$

$$\pm \frac{n(n-1)(n-2)x^3}{3!} + \dots \quad [x^2 < 1]$$

$$(1 \pm x)^{-n} = 1 \mp nx + \frac{n(n+1)}{2!}x^2$$

$$\mp \frac{n(n+1)(n+2)}{3!}x^3 + \dots \quad [x^2 < 1]$$

$$(1 \pm x)^{\frac{1}{2}} = 1 \pm \frac{1}{2}x - \frac{1}{2 \cdot 4}x^2 \pm \frac{1 \cdot 3}{2 \cdot 4 \cdot 6}x^3$$

$$- \frac{1 \cdot 3 \cdot 5}{2 \cdot 4 \cdot 6 \cdot 8}x^4 \pm \ldots \qquad [x^2 < 1]$$

$$(1 \pm x)^{-\frac{1}{2}} = 1 \mp \frac{1}{2}x + \frac{1 \cdot 3}{2 \cdot 4}x^2 \mp \frac{1 \cdot 3 \cdot 5}{2 \cdot 4 \cdot 6}x^3$$

$$+ \frac{1 \cdot 3 \cdot 5 \cdot 7}{2 \cdot 4 \cdot 6 \cdot 8}x^4 \mp \ldots \qquad [x^2 < 1]$$

$$(1 \pm x^2)^{\frac{1}{2}} = 1 \pm \frac{1}{2}x^2 - \frac{x^4}{2 \cdot 4} \pm \frac{1 \cdot 3}{2 \cdot 4 \cdot 6}x^6$$

$$- \frac{1 \cdot 3 \cdot 5}{2 \cdot 4 \cdot 6 \cdot 8}x^8 \pm \ldots \qquad [x^2 < 1]$$

$$(1 \pm x)^{-1} = 1 \mp x + x^2 \mp x^3 + x^4 \mp x^5 + \ldots$$

$$[x^2 < 1]$$

$$(1 \pm x)^{-2} = 1 \mp 2x + 3x^2 \mp 4x^3 + 5x^4 \mp \ldots$$

$$[x^2 < 1]$$

$$e^x = 1 + x + \frac{x^2}{2!} + \frac{x^3}{3!} + \frac{x^4}{4!} + \ldots$$

$$e^{-x^2} = 1 - x^2 + \frac{x^4}{2!} - \frac{x^6}{3!} + \frac{x^8}{4!} - \ldots$$

$$a^x = 1 + x \log a + \frac{(x \log a)^2}{2!} + \frac{(x \log a)^3}{3!} + \ldots$$

$$\log x = (x-1) - \frac{1}{2}(x-1)^2 + \frac{1}{3}(x-1)^3 - \ldots$$

$$[0 < x < 2]$$

$$\log x = \frac{x-1}{x} + \frac{1}{2}\left(\frac{x-1}{x}\right)^2 + \frac{1}{3}\left(\frac{x-1}{x}\right)^3 + \ldots$$

$$\left[x > \frac{1}{2}\right]$$

$$\log x = 2\left[\left(\frac{x-1}{x+1}\right) + \frac{1}{3}\left(\frac{x-1}{x+1}\right)^3 + \frac{1}{5}\left(\frac{x-1}{x+1}\right)^5 + \ldots\right]$$

$$[x > 0]$$

$$\log(1+x) = x - \frac{1}{2}x^2 + \frac{1}{3}x^3 - \frac{1}{4}x^4 + \ldots$$

$$[x^2 < 1]$$

$$\log\left(\frac{1+x}{1-x}\right) = 2\left[x + \frac{1}{3}x^3 + \frac{1}{5}x^5 + \frac{1}{7}x^7 + \ldots\right]$$

$$[x^2 < 1]$$

$$\log\left(\frac{x+1}{x-1}\right) = 2\left[\frac{1}{x} + \frac{1}{3}\left(\frac{1}{x}\right)^3 + \frac{1}{5}\left(\frac{1}{x}\right)^5 + \ldots\right]$$

$$[x^2 > 1]$$

$$\sin x = x - \frac{x^3}{3!} + \frac{x^5}{5!} - \frac{x^7}{7!} + \ldots$$

$$\cos x = 1 - \frac{x^2}{2!} + \frac{x^4}{4!} - \frac{x^6}{6!} + \ldots$$

$$\tan x = x + \frac{x^3}{3} + \frac{2x^5}{15} + \frac{17x^7}{315}$$

$$+ \ldots + \frac{2^{2n}(2^{2n}-1)B_{2n-1}x^{2n-1}}{(2n)!}$$

$$\left[x^2 < \frac{\pi^2}{4} \right]$$

$$\operatorname{ctn} x = \frac{1}{x} - \frac{x}{3} - \frac{x^3}{45} - \frac{2x^5}{945}$$

$$- \ldots - \frac{B_{2n-1}(2x)^{2n}}{(2n)!x} - \ldots$$

$$[x^2 < \pi^2]$$

$$\sec x = 1 + \frac{x^2}{2!} + \frac{5x^4}{4!} + \frac{61x^6}{6!} + \ldots$$

$$+ \frac{B_{2n}x^{2n}}{(2n)!} + \ldots \qquad \left[x^2 < \frac{\pi^2}{4} \right]$$

$$\csc x = \frac{1}{x} + \frac{x}{3!} + \frac{7x^3}{3\cdot 5!} + \frac{31x^5}{3\cdot 7!}$$

$$+ \ldots + \frac{2(2^{2n+1}-1)}{(2n+2)!}B_{2n+1}x^{2n+1} + \ldots$$

$$[x^2 < \pi^2]$$

$$\sin^{-1} x = x + \frac{x^3}{6} + \frac{(1\cdot 3)x^5}{(2\cdot 4)5} + \frac{(1\cdot 3\cdot 5)x^7}{(2\cdot 4\cdot 6)7} + \ldots$$

$$[x^2 < 1]$$

$$\tan^{-1} x = x - \frac{1}{3}x^3 + \frac{1}{5}x^5 - \frac{1}{7}x^7 + \dots$$

$$[x^2 < 1]$$

$$\sec^{-1} x = \frac{\pi}{2} - \frac{1}{x} - \frac{1}{6x^3}$$

$$- \frac{1\cdot 3}{(2\cdot 4)5x^5} - \frac{1\cdot 3\cdot 5}{(2\cdot 4\cdot 6)7x^7} - \dots$$

$$[x^2 > 1]$$

$$\sinh x = x + \frac{x^3}{3!} + \frac{x^5}{5!} + \frac{x^7}{7!} + \dots$$

$$\cosh x = 1 + \frac{x^2}{2!} + \frac{x^4}{4!} + \frac{x^6}{6!} + \frac{x^8}{8!} + \dots$$

$$\tanh x = (2^2 - 1)2^2 B_1 \frac{x}{2!} - (2^4 - 1)2^4 B_3 \frac{x^3}{4!}$$

$$+ (2^6 - 1)2^6 B_5 \frac{x^5}{6!} - \dots \qquad \left[x^2 < \frac{\pi^2}{4} \right]$$

$$\operatorname{ctnh} x = \frac{1}{x}\left(1 + \frac{2^2 B_1 x^2}{2!} - \frac{2^4 B_3 x^4}{4!} \right.$$

$$\left. + \frac{2^6 B_5 x^6}{6!} - \dots \right)$$

$$[x^2 < \pi^2]$$

$$\operatorname{sech} x = 1 - \frac{B_2 x^2}{2!} + \frac{B_4 x^4}{4!} - \frac{B_6 x^6}{6!} + \dots$$

$$\left[x^2 < \frac{\pi^2}{4} \right]$$

$$\operatorname{csch} x = \frac{1}{x} - (2-1)2B_1 \frac{x}{2!}$$

$$+ (2^3 - 1)2B_3 \frac{x^3}{4!} - \dots$$

$$[x^2 < \pi^2]$$

$$\sinh^{-1} x = x - \frac{1}{2} \frac{x^3}{3} + \frac{1 \cdot 3}{2 \cdot 4} \frac{x^5}{5} - \frac{1 \cdot 3 \cdot 5}{2 \cdot 4 \cdot 6} \frac{x^7}{7} + \dots$$

$$[x^2 < 1]$$

$$\tanh^{-1} x = x + \frac{x^3}{3} + \frac{x^5}{5} + \frac{x^7}{7} + \dots \qquad [x^2 < 1]$$

$$\operatorname{ctnh}^{-1} x = \frac{1}{x} + \frac{1}{3x^3} + \frac{1}{5x^5} + \dots \qquad [x^2 > 1]$$

$$\operatorname{csch}^{-1} x = \frac{1}{x} - \frac{1}{2 \cdot 3x^3} + \frac{1 \cdot 3}{2 \cdot 4 \cdot 5x^5}$$

$$- \frac{1 \cdot 3 \cdot 5}{2 \cdot 4 \cdot 6 \cdot 7x^7} + \dots \qquad [x^2 > 1]$$

$$\int_0^x e^{-t^2} dt = x - \frac{1}{3}x^3 + \frac{x^5}{5 \cdot 2!} - \frac{x^7}{7 \cdot 3!} + \dots$$

3. Error Function

The following function, known as the error function, erf x, arises frequently in applications:

$$\text{erf } x = \frac{2}{\sqrt{\pi}} \int_0^x e^{-t^2} dt$$

The integral cannot be represented in terms of a finite number of elementary functions, therefore values of erf x have been compiled in tables. The following is the series for erf x:

$$\text{erf } x = \frac{2}{\sqrt{\pi}} \left[x - \frac{x^3}{3} + \frac{x^5}{5 \cdot 2!} - \frac{x^7}{7 \cdot 3!} + \ldots \right]$$

There is a close relation between this function and the area under the standard normal curve. For evaluation it is convenient to use z instead of x; then erf z may be evaluated from the area $F(z)$ by use of the relation

$$\text{erf } z = 2F(\sqrt{2}\,z)$$

Example

$$\text{erf}(0.5) = 2F[(1.414)(0.5)] = 2F(0.707)$$

By interpolation, $F(0.707) = 0.260$; thus, erf$(0.5) = 0.520$.

6 Differential Calculus

1. Notation

For the following equations, the symbols $f(x)$, $g(x)$, etc., represent functions of x. The value of a function $f(x)$ at $x=a$ is denoted $f(a)$. For the function $y=f(x)$ the derivative of y with respect to x is denoted by one of the following:

$$\frac{dy}{dx}, \quad f'(x), \quad D_x y, \quad y'.$$

Higher derivatives are as follows:

$$\frac{d^2 y}{dx^2} = \frac{d}{dx}\left(\frac{dy}{dx}\right) = \frac{d}{dx} f'(x) = f''(x)$$

$$\frac{d^3 y}{dx^3} = \frac{d}{dx}\left(\frac{d^2 y}{dx^2}\right) = \frac{d}{dx} f''(x) = f'''(x), \text{ etc.}$$

and values of these at $x=a$ are denoted $f''(a)$, $f'''(a)$, etc. (see Table of Derivatives).

2. Slope of a Curve

The tangent line at a point $P(x, y)$ of the curve $y=f(x)$ has a slope $f'(x)$ provided that $f'(x)$ exists at P. The slope at P is defined to be that of the tangent line at P. The tangent line at $P(x_1, y_1)$ is given by

$$y - y_1 = f'(x_1)(x - x_1).$$

The *normal line* to the curve at $P(x_1, y_1)$ has slope $-1/f'(x_1)$ and thus obeys the equation

$$y - y_1 = [-1/f'(x_1)](x - x_1)$$

(The slope of a vertical line is not defined.)

3. Angle of Intersection of Two Curves

Two curves, $y = f_1(x)$ and $y = f_2(x)$, that intersect at a point $P(X, Y)$ where derivatives $f_1'(X)$, $f_2'(X)$ exist, have an angle (α) of intersection given by

$$\tan \alpha = \frac{f_2'(X) - f_1'(X)}{1 + f_2'(X) \cdot f_1'(X)}.$$

If $\tan \alpha > 0$, then α is the acute angle; if $\tan \alpha < 0$, then α is the obtuse angle.

4. Radius of Curvature

The radius of curvature R of the curve $y = f(x)$ at point $P(x, y)$ is

$$R = \frac{\{1 + [f'(x)]^2\}^{3/2}}{f''(x)}$$

In polar coordinates (θ, r) the corresponding formula is

$$R = \frac{\left[r^2 + \left(\dfrac{dr}{d\theta}\right)^2\right]^{3/2}}{r^2 + 2\left(\dfrac{dr}{d\theta}\right)^2 - r\dfrac{d^2r}{d\theta^2}}$$

The *curvature K* is $1/R$.

5. Relative Maxima and Minima

The function f has a relative maximum at $x = a$ if $f(a) \geq f(a + c)$ for all values of c (positive or negative) that are sufficiently near zero. The function f has a relative minimum at $x = b$ if $f(b) \leq f(b + c)$ for all values of c that are sufficiently close to zero. If the function f is defined on the closed interval $x_1 \leq x \leq x_2$, and has a relative maximum or minimum at $x = a$, where $x_1 < a < x_2$, and if the derivative $f'(x)$ exists at $x = a$, then $f'(a) = 0$. It is noteworthy that a relative maximum or minimum may occur at a point where the derivative does not exist. Further, the derivative may vanish at a point that is neither a maximum nor a minimum for the function. Values of x for which $f'(x) = 0$ are called "critical values." To determine whether a critical value of x, say x_c, is a relative maximum or minimum for the function at x_c, one may use the second derivative test

1. If $f''(x_c)$ is positive, $f(x_c)$ is a minimum

2. If $f''(x_c)$ is negative, $f(x_c)$ is a maximum

3. If $f''(x_c)$ is zero, no conclusion may be made

The sign of the derivative as x advances through x_c may also be used as a test. If $f'(x)$ changes from positive to zero to negative, then a maximum occurs at x_c, whereas a change in $f'(x)$ from negative to zero to positive indicates a minimum. If $f'(x)$ does not change sign as x advances through x_c, then the point is neither a maximum nor a minimum.

6. Points of Inflection of a Curve

The sign of the second derivative of f indicates whether the graph of $y = f(x)$ is concave upward or concave downward:

$$f''(x) > 0: \text{concave upward}$$

$$f''(x) < 0: \text{concave downward}$$

A point of the curve at which the direction of concavity changes is called a point of inflection (Figure 6.1). Such a point may occur where $f''(x) = 0$ or where $f''(x)$ becomes infinite. More precisely, if the function $y = f(x)$ and its first derivative $y' = f'(x)$ are continuous in the interval $a \le x \le b$, and if $y'' = f''(x)$ exists in $a < x < b$, then the graph of $y = f(x)$ for $a < x < b$ is concave

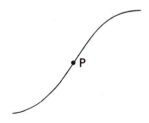

FIGURE 6.1. Point of inflection.

upward if $f''(x)$ is positive and concave downward if $f''(x)$ is negative.

7. Taylor's Formula

If f is a function that is continuous on an interval that contains a and x, and if its first $(n+1)$ derivatives are continuous on this interval, then

$$f(x) = f(a) + f'(a)(x-a) + \frac{f''(a)}{2!}(x-a)^2$$

$$+ \frac{f'''(a)}{3!}(x-a)^3 + \ldots$$

$$+ \frac{f^{(n)}(a)}{n!}(x-a)^n + R,$$

where R is called the *remainder*. There are various common forms of the remainder:

Lagrange's form:

$$R = f^{(n+1)}(\beta) \cdot \frac{(x-a)^{n+1}}{(n+1)!}; \; \beta \text{ between } a \text{ and } x.$$

Cauchy's form:

$$R = f^{(n+1)}(\beta) \cdot \frac{(x-\beta)^n (x-a)}{n!};$$

$$\beta \text{ between } a \text{ and } x.$$

Integral form:

$$R = \int_a^x \frac{(x-t)^n}{n!} f^{(n+1)}(t)\, dt.$$

8. Indeterminant Forms

If $f(x)$ and $g(x)$ are continuous in an interval that includes $x = a$ and if $f(a) = 0$ and $g(a) = 0$, the limit $\lim_{x \to a}(f(x)/g(x))$ takes the form "0/0", called an *indeterminant form*. *L'Hôpital's rule* is

$$\lim_{x \to a} \frac{f(x)}{g(x)} = \lim_{x \to a} \frac{f'(x)}{g'(x)}.$$

Similarly, it may be shown that if $f(x) \to \infty$ and $g(x) \to \infty$ as $x \to a$, then

$$\lim_{x \to a} \frac{f(x)}{g(x)} = \lim_{x \to a} \frac{f'(x)}{g'(x)}.$$

(The above holds for $x \to \infty$.)

Examples

$$\lim_{x \to 0} \frac{\sin x}{x} = \lim_{x \to 0} \frac{\cos x}{1} = 1$$

$$\lim_{x \to \infty} \frac{x^2}{e^x} = \lim_{x \to \infty} \frac{2x}{e^x} = \lim_{x \to \infty} \frac{2}{e^x} = 0$$

9. Numerical Methods

a. *Newton's method* for approximating roots of the equation $f(x) = 0$: A first estimate x_1 of the root is made; then provided that $f'(x_1) \neq 0$, a better approximation is x_2

$$x_2 = x_1 - \frac{f(x_1)}{f'(x_1)}.$$

The process may be repeated to yield a third approximation x_3 to the root:

$$x_3 = x_2 - \frac{f(x_2)}{f'(x_2)}$$

provided $f'(x_2)$ exists. The process may be repeated. (In certain rare cases the process will not converge.)

b. *Trapezoidal rule for areas* (Figure 6.2): For the function $y = f(x)$ defined on the interval (a, b) and positive there, take n equal subintervals of width $\Delta x = (b-a)/n$. The area bounded by the curve between

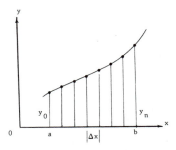

FIGURE 6.2. Trapezoidal rule for area.

$x = a$ and $x = b$ (or definite integral of $f(x)$) is approximately the sum of trapezoidal areas, or

$$A \sim \left(\frac{1}{2}y_0 + y_1 + y_2 + \ldots + y_{n-1} + \frac{1}{2}y_n \right)(\Delta x)$$

Estimation of the error (E) is possible if the second derivative can be obtained:

$$E = \frac{b-a}{12} f''(c)(\Delta x)^2,$$

where c is some number between a and b.

10. Functions of Two Variables

For the function of two variables, denoted $z = f(x, y)$, if y is held constant, say at $y = y_1$, then the resulting function is a function of x only. Similarly, x may be held constant at x_1, to give the resulting function of y.

- ### The Gas Laws

A familiar example is afforded by the ideal gas law that relates the pressure p, the volume V and the absolute temperature T of an ideal gas:

$$pV = nRT$$

where n is the number of moles and R is the gas constant per mole, 8.31 $(\text{J} \cdot {}^\circ\text{K}^{-1} \cdot \text{mole}^{-1})$. By rearrangement, any one of the three variables may be expressed as a function of the other two. Further, either one of these two may be held constant. If T is

70

held constant, then we get the form known as Boyle's law:

$$p = kV^{-1} \qquad \text{(Boyle's law)}$$

where we have denoted nRT by the constant k and, of course, $V > 0$. If the pressure remains constant, we have Charles' law:

$$V = bT \qquad \text{(Charles' law)}$$

where the constant b denotes nR/p. Similarly, volume may be kept constant:

$$p = aT$$

where now the constant, denoted a, is nR/V.

11. Partial Derivatives

The physical example afforded by the ideal gas law permits clear interpretations of processes in which one of the variables is held constant. More generally, we may consider a function $z = f(x, y)$ defined over some region of the x–y-plane in which we hold one of the two coordinates, say y, constant. If the resulting function of x is differentiable at a point (x, y) we denote this derivative by one of the notations

$$f_x, \qquad \delta f/dx, \qquad \delta z/dx$$

called the *partial derivative with respect to x*. Similarly, if x is held constant and the resulting function of y is differentiable, we get the *partial derivative with respect to y*, denoted by one of the following:

$$f_y \qquad \delta f/dy \qquad \delta z/dy$$

71

Example

Given $z = x^4 y^3 - y \sin x + 4y$, then

$\delta z / dx = 4(xy)^3 - y \cos x;$

$\delta z / dy = 3x^4 y^2 - \sin x + 4.$

7 Integral Calculus

1. Indefinite Integral

If $F(x)$ is differentiable for all values of x in the interval (a, b) and satisfies the equation $dy/dx = f(x)$, then $F(x)$ is an integral of $f(x)$ with respect to x. The notation is $F(x) = \int f(x)\, dx$ or, in differential form, $dF(x) = f(x)\, dx$.

For any function $F(x)$ that is an integral of $f(x)$ it follows that $F(x) + C$ is also an integral. We thus write

$$\int f(x)\, dx = F(x) + C.$$

(See Table of Integrals.)

2. Definite Integral

Let $f(x)$ be defined on the interval $[a, b]$ which is partitioned by points $x_1, x_2, \ldots, x_j, \ldots, x_{n-1}$ between $a = x_0$ and $b = x_n$. The jth interval has length $\Delta x_j = x_j - x_{j-1}$, which may vary with j. The sum $\sum_{j=1}^{n} f(v_j) \Delta x_j$, where v_j is arbitrarily chosen in the jth subinterval, depends on the numbers $x_0, \ldots, x_n$ and the choice of the v as well as f; but if such sums approach a common value as all Δx approach zero, then this value is the definite integral of f over the interval (a, b) and is denoted $\int_a^b f(x)\,dx$. The *fundamental theorem of integral calculus* states that

$$\int_a^b f(x)\,dx = F(b) - F(a),$$

where F is any continuous indefinite integral of f in the interval (a, b).

3. Properties

$$\int_a^b [f_1(x) + f_2(x) + \cdots + f_j(x)]\,dx = \int_a^b f_1(x)\,dx$$

$$+ \int_a^b f_2(x)\,dx + \cdots + \int_a^b f_j(x)\,dx.$$

$$\int_a^b cf(x)\,dx = c\int_a^b f(x)\,dx, \quad \text{if } c \text{ is a constant.}$$

$$\int_a^b f(x)\,dx = -\int_b^a f(x)\,dx.$$

$$\int_a^b f(x)\,dx = \int_a^c f(x)\,dx + \int_c^b f(x)\,dx.$$

4. Common Applications of the Definite Integral

- *Area (Rectangular Coordinates)*

  Given the function $y = f(x)$ such that $y > 0$ for all x between a and b, the area bounded by the curve

$y = f(x)$, the x-axis, and the vertical lines $x = a$ and $x = b$ is

$$A = \int_a^b f(x)\, dx.$$

- *Length of Arc (Rectangular Coordinates)*

  Given the smooth curve $f(x, y) = 0$ from point (x_1, y_1) to point (x_2, y_2), the length between these points is

  $$L = \int_{x_1}^{x_2} \sqrt{1 + (dy/dx)^2}\, dx,$$

  $$L = \int_{y_1}^{y_2} \sqrt{1 + (dx/dy)^2}\, dy.$$

- *Mean Value of a Function*

  The mean value of a function $f(x)$ continuous on $[a, b]$ is

  $$\frac{1}{(b-a)} \int_a^b f(x)\, dx.$$

- *Area (Polar Coordinates)*

  Given the curve $r = f(\theta)$, continuous and non-negative for $\theta_1 \le \theta \le \theta_2$, the area enclosed by this curve and the radial lines $\theta = \theta_1$ and $\theta = \theta_2$ is given by

  $$A = \int_{\theta_1}^{\theta_2} \frac{1}{2} [f(\theta)]^2\, d\theta.$$

- *Length of Arc (Polar Coordinates)*

  Given the curve $r = f(\theta)$ with continuous derivative $f'(\theta)$ on $\theta_1 \le \theta \le \theta_2$, the length of arc from $\theta = \theta_1$ to $\theta = \theta_2$ is

  $$L = \int_{\theta_1}^{\theta_2} \sqrt{[f(\theta)]^2 + [f'(\theta)]^2}\, d\theta.$$

- *Volume of Revolution*

  Given a function $y = f(x)$ continuous and non-negative on the interval (a, b), when the region bounded by $f(x)$ between a and b is revolved about the x-axis the volume of revolution is

  $$V = \pi \int_a^b [f(x)]^2\, dx.$$

- *Surface Area of Revolution*
  (revolution about the x-axis, between a and b)

  If the portion of the curve $y = f(x)$ between $x = a$ and $x = b$ is revolved about the x-axis, the area A of the surface generated is given by the following:

  $$A = \int_a^b 2\pi f(x)\{1 + [f'(x)]^2\}^{1/2}\, dx$$

- *Work*

  If a variable force $f(x)$ is applied to an object in the direction of motion along the x-axis between $x = a$ and $x = b$, the work done is

$$W = \int_a^b f(x)\, dx.$$

5. Cylindrical and Spherical Coordinates

a. Cylindrical coordinates (Figure 7.1)

$$x = r \cos \theta$$
$$y = r \sin \theta$$

element of volume $dV = r\, dr\, d\theta\, dz$.

b. Spherical coordinates (Figure 7.2)

$$x = \rho \sin \phi \cos \theta$$
$$y = \rho \sin \phi \sin \theta$$
$$z = \rho \cos \phi$$

element of volume $dV = \rho^2 \sin \phi\, d\rho, d\phi\, d\theta$.

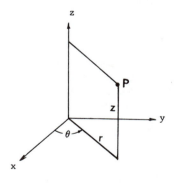

FIGURE 7.1. Cylindrical coordinates.

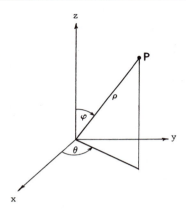

FIGURE 7.2. Spherical coordinates.

6. Double Integration

The evaluation of a double integral of $f(x, y)$ over a plane region R

$$\iint_R f(x, y)\, dA$$

is practically accomplished by iterated (repeated) integration. For example, suppose that a vertical straight line meets the boundary of R in at most two points so that there is an upper boundary, $y = y_2(x)$, and a lower boundary, $y = y_1(x)$. Also, it is assumed that these functions are continuous from a to b. (See Figure 7.3). Then

$$\iint_R f(x, y)\, dA = \int_a^b \left(\int_{y_1(x)}^{y_2(x)} f(x, y)\, dy \right) dx$$

78

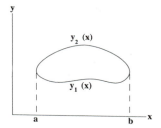

FIGURE 7.3. Region R bounded by $y_2(x)$ and $y_1(x)$.

If R has left-hand boundary, $x = x_1(y)$, and a right-hand boundary, $x = x_2(y)$, which are continuous from c to d (the extreme values of y in R) then

$$\iint_R f(x,y)\, dA = \int_c^d \left(\int_{x_1(y)}^{x_2(y)} f(x,y)\, dx \right) dy$$

Such integrations are sometimes more convenient in polar coordinates, $x = r\cos\theta$, $y = r\sin\theta$; $dA = r\,dr\,d\theta$.

7. Surface Area and Volume by Double Integration

For the surface given by $z = f(x,y)$, which projects onto the closed region R of the x-y-plane, one may calculate the volume V bounded above by the surface and below by R, and the surface area S by the following:

$$V = \iint_R z\, dA = \iint_R f(x,y)\, dx\, dy$$

$$S = \iint_R \left[1 + (\delta z/\delta x)^2 + (\delta z/\delta y)^2\right]^{1/2} dx\, dy$$

[In polar coordinates, (r, θ), we replace dA by $r\,dr\,d\theta$].

8. Centroid

The centroid of a region R of the x-y-plane is a point (x', y') where

$$x' = \frac{1}{A} \iint_R x\,dA; \qquad y' = \frac{1}{A} \iint_R y\,dA$$

and A is the area of the region.

Example

For the circular sector of angle 2α and radius R, the area A is αR^2; the integral needed for x', expressed in polar coordinates is

$$\iint x\,dA = \int_{-\alpha}^{\alpha}\int_0^R (r\cos\theta)r\,dr\,d\theta$$

$$= \left[\frac{R^3}{3}\sin\theta\right]_{-\alpha}^{+\alpha} = \frac{2}{3}R^3\sin\alpha$$

and thus,

$$x' = \frac{\dfrac{2}{3}R^3\sin\alpha}{\alpha R^2} = \frac{2}{3}R\,\frac{\sin\alpha}{\alpha}.$$

Centroids of some common regions are shown below:

Centroids

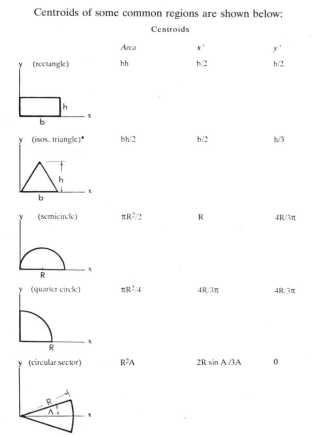

	Area	x'	y'
y (rectangle)	bh	b/2	h/2
y (isos. triangle)*	bh/2	b/2	h/3
y (semicircle)	πR²/2	R	4R/3π
y (quarter circle)	πR²/4	4R/3π	4R/3π
y (circular sector)	R²A	2R sin A /3A	0

* y' = h/3 for any triangle of altitude h.

FIGURE 7.4.

81

8 Vector Analysis

1. Vectors

Given the set of mutually perpendicular unit vectors $\mathbf{i}$, $\mathbf{j}$, and $\mathbf{k}$ (Figure 8.1), then any vector in the space may be represented as $\mathbf{F} = a\mathbf{i} + b\mathbf{j} + c\mathbf{k}$, where a, b, and c are *components*.

- *Magnitude of $\mathbf{F}$*

$$|\mathbf{F}| = (a^2 + b^2 + c^2)^{\frac{1}{2}}$$

- *Product by scalar p*

$$p\mathbf{F} = pa\mathbf{i} + pb\mathbf{j} + pc\mathbf{k}.$$

- *Sum of $\mathbf{F}_1$ and $\mathbf{F}_2$*

$$\mathbf{F}_1 + \mathbf{F}_2 = (a_1 + a_2)\mathbf{i} + (b_1 + b_2)\mathbf{j} + (c_1 + c_2)\mathbf{k}$$

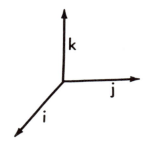

FIGURE 8.1. The unit vectors $\mathbf{i}$, $\mathbf{j}$, and $\mathbf{k}$.

- *Scalar Product*

$$\mathbf{F}_1 \cdot \mathbf{F}_2 = a_1 a_2 + b_1 b_2 + c_1 c_2$$

(Thus, $\mathbf{i} \cdot \mathbf{i} = \mathbf{j} \cdot \mathbf{j} = \mathbf{k} \cdot \mathbf{k} = 1$ and $\mathbf{i} \cdot \mathbf{j} = \mathbf{j} \cdot \mathbf{k} = \mathbf{k} \cdot \mathbf{i} = 0$.)
Also

$$\mathbf{F}_1 \cdot \mathbf{F}_2 = \mathbf{F}_2 \cdot \mathbf{F}_1$$

$$(\mathbf{F}_1 + \mathbf{F}_2) \cdot \mathbf{F}_3 = \mathbf{F}_1 \cdot \mathbf{F}_3 + \mathbf{F}_2 \cdot \mathbf{F}_3$$

- *Vector Product*

$$\mathbf{F}_1 \times \mathbf{F}_2 = \begin{vmatrix} \mathbf{i} & \mathbf{j} & \mathbf{k} \\ a_1 & b_1 & c_1 \\ a_2 & b_2 & c_2 \end{vmatrix}$$

(Thus, $\mathbf{i} \times \mathbf{i} = \mathbf{j} \times \mathbf{j} = \mathbf{k} \times \mathbf{k} = 0$, $\mathbf{i} \times \mathbf{j} = \mathbf{k}$, $\mathbf{j} \times \mathbf{k} = \mathbf{i}$, and $\mathbf{k} \times \mathbf{i} = \mathbf{j}$.)
Also,

$$\mathbf{F}_1 \times \mathbf{F}_2 = -\mathbf{F}_2 \times \mathbf{F}_1$$

$$(\mathbf{F}_1 + \mathbf{F}_2) \times \mathbf{F}_3 = \mathbf{F}_1 \times \mathbf{F}_3 + \mathbf{F}_2 \times \mathbf{F}_3$$

$$\mathbf{F}_1 \times (\mathbf{F}_2 + \mathbf{F}_3) = \mathbf{F}_1 \times \mathbf{F}_2 + \mathbf{F}_1 \times \mathbf{F}_3$$

$$\mathbf{F}_1 \times (\mathbf{F}_2 \times \mathbf{F}_3) = (\mathbf{F}_1 \cdot \mathbf{F}_3)\mathbf{F}_2 - (\mathbf{F}_1 \cdot \mathbf{F}_2)\mathbf{F}_3$$

$$\mathbf{F}_1 \cdot (\mathbf{F}_2 \times \mathbf{F}_3) = (\mathbf{F}_1 \times \mathbf{F}_2) \cdot \mathbf{F}_3$$

2. *Vector Differentiation*

If $\mathbf{V}$ is a vector function of a scalar variable t, then

$$\mathbf{V} = a(t)\mathbf{i} + b(t)\mathbf{j} + c(t)\mathbf{k}$$

and

$$\frac{d\mathbf{V}}{dt} = \frac{da}{dt}\mathbf{i} + \frac{db}{dt}\mathbf{j} + \frac{dc}{dt}\mathbf{k}.$$

For several vector functions $\mathbf{V}_1, \mathbf{V}_2, \ldots, \mathbf{V}_n$

$$\frac{d}{dt}(\mathbf{V}_1 + \mathbf{V}_2 + \ldots + \mathbf{V}_n) = \frac{d\mathbf{V}_1}{dt} + \frac{d\mathbf{V}_2}{dt} + \ldots + \frac{d\mathbf{V}_n}{dt},$$

$$\frac{d}{dt}(\mathbf{V}_1 \bullet \mathbf{V}_2) = \frac{d\mathbf{V}_1}{dt} \bullet \mathbf{V}_2 + \mathbf{V}_1 \bullet \frac{d\mathbf{V}_2}{dt},$$

$$\frac{d}{dt}(\mathbf{V}_1 \times \mathbf{V}_2) = \frac{d\mathbf{V}_1}{dt} \times \mathbf{V}_2 + \mathbf{V}_1 \times \frac{d\mathbf{V}_2}{dt}.$$

For a scalar valued function $g(x, y, z)$

(**gradient**) $\operatorname{grad} g = \nabla g = \dfrac{\delta g}{\delta x}\mathbf{i} + \dfrac{\delta g}{\delta y}\mathbf{j} + \dfrac{\delta g}{\delta z}\mathbf{k}.$

For a vector valued function $\mathbf{V}(a, b, c)$, where a, b, c are each a function of x, y, and z,

(**divergence**) $\operatorname{div} \mathbf{V} = \nabla \bullet \mathbf{V} = \dfrac{\delta a}{\delta x} + \dfrac{\delta b}{\delta y} + \dfrac{\delta c}{\delta z}$

(**curl**) $\operatorname{curl} \mathbf{V} = \nabla \times \mathbf{V} = \begin{vmatrix} \mathbf{i} & \mathbf{j} & \mathbf{k} \\ \dfrac{\delta}{\delta x} & \dfrac{\delta}{\delta y} & \dfrac{\delta}{\delta z} \\ a & b & c \end{vmatrix}$

Also,

$$\text{div} \, \text{grad} \, g = \nabla^2 g = \frac{\delta^2 g}{\delta x^2} + \frac{\delta^2 g}{\delta y^2} + \frac{\delta^2 g}{\delta z^2}$$

and

$$\text{curl} \, \text{grad} \, g = \mathbf{0}; \qquad \text{div} \, \text{curl} \, \mathbf{V} = 0;$$

$$\text{curl} \, \text{curl} \, \mathbf{V} = \text{grad} \, \text{div} \, \mathbf{V} - (\mathbf{i} \nabla^2 a + \mathbf{j} \nabla^2 b + \mathbf{k} \nabla^2 c).$$

3. Divergence Theorem (Gauss)

Given a vector function F with continuous partial derivatives in a region R bounded by a closed surface S, then

$$\iiint_R \text{div} \, \mathbf{F} \, dV = \iint_S \mathbf{n} \cdot \mathbf{F} \, dS,$$

where $\mathbf{n}$ is the (sectionally continuous) unit normal to S.

4. Stokes' Theorem

Given a vector function with continuous gradient over a surface S that consists of portions that are piecewise smooth and bounded by regular closed curves such as C, then

$$\iint_S \mathbf{n} \cdot \text{curl} \, \mathbf{F} \, dS = \oint_C \mathbf{F} \cdot d\mathbf{r}$$

5. Planar Motion in Polar Coordinates

Motion in a plane may be expressed with regard to polar coordinates (r, θ). Denoting the position vector by $\mathbf{r}$ and its magnitude by r, we have $\mathbf{r} = r\mathbf{R}(\theta)$, where $\mathbf{R}$ is the unit vector. Also, $d\mathbf{R}/d\theta = \mathbf{P}$, a unit vector perpendicular to **R**. The velocity and acceleration are then

$$\mathbf{v} = \frac{dr}{dt}\mathbf{R} + r\frac{d\theta}{dt}\mathbf{P};$$

$$\mathbf{a} = \left[\frac{d^2r}{dt^2} - r\left(\frac{d\theta}{dt}\right)^2\right]\mathbf{R} + \left[r\frac{d^2\theta}{dt^2} + 2\frac{dr}{dt}\frac{d\theta}{dt}\right]\mathbf{P}.$$

Note that the component of acceleration in the **P** direction (transverse component) may also be written

$$\frac{1}{r}\frac{d}{dt}\left(r^2\frac{d\theta}{dt}\right)$$

so that in purely radial motion it is zero and

$$r^2\frac{d\theta}{dt} = C \text{ (constant)}$$

which means that the position vector sweeps out area at a constant rate (see Area in Polar Coordinates, Section 7.4).

9 Special Functions

1. Hyperbolic Functions

$$\sinh x = \frac{e^x - e^{-x}}{2}$$

$$\operatorname{csch} x = \frac{1}{\sinh x}$$

$$\cosh x = \frac{e^x + e^{-x}}{2}$$

$$\operatorname{sech} x = \frac{1}{\cosh x}$$

$$\tanh x = \frac{e^x - e^{-x}}{e^x + e^{-x}}$$

$$\operatorname{ctnh} x = \frac{1}{\tanh x}$$

$$\sinh(-x) = -\sinh x$$

$$\operatorname{ctnh}(-x) = -\operatorname{ctnh} x$$

$$\cosh(-x) = \cosh x$$

$$\operatorname{sech}(-x) = \operatorname{sech} x$$

$$\tanh(-x) = -\tanh x$$

$$\operatorname{csch}(-x) = -\operatorname{csch} x$$

$$\tanh x = \frac{\sinh x}{\cosh x}$$

$$\operatorname{ctnh} x = \frac{\cosh x}{\sinh x}$$

$$\cosh^2 x - \sinh^2 x = 1$$

$$\cosh^2 x = \frac{1}{2}(\cosh 2x + 1)$$

$$\sinh^2 x = \frac{1}{2}(\cosh 2x - 1)$$

$$\operatorname{ctnh}^2 x - \operatorname{csch}^2 x = 1$$

$$\operatorname{csch}^2 x - \operatorname{sech}^2 x = \operatorname{csch}^2 x \operatorname{sech}^2 x$$

$$\tanh^2 x + \operatorname{sech}^2 x = 1$$

$$\sinh(x+y) = \sinh x \cosh y + \cosh x \sinh y$$

$$\cosh(x+y) = \cosh x \cosh y + \sinh x \sinh y$$

$$\sinh(x-y) = \sinh x \cosh y - \cosh x \sinh y$$

$$\cosh(x-y) = \cosh x \cosh y - \sinh x \sinh y$$

$$\tanh(x+y) = \frac{\tanh x + \tanh y}{1 + \tanh x \tanh y}$$

$$\tanh(x-y) = \frac{\tanh x - \tanh y}{1 - \tanh x \tanh y}$$

2. Gamma Function (Generalized Factorial Function)

The gamma function, denoted $\Gamma(x)$, is defined by

$$\Gamma(x) = \int_0^\infty e^{-t} t^{x-1}\, dt \qquad (x > 0)$$

- Properties

$$\Gamma(x+1) = x\Gamma(x) \qquad\qquad (x > 0)$$

$$\Gamma(1) = 1$$

$$\Gamma(n+1) = n\Gamma(n) = n! \qquad\qquad (n = 1, 2, 3, \ldots)$$

$$\Gamma(x)\Gamma(1-x) = \pi/\sin \pi x$$

$$\Gamma\left(\frac{1}{2}\right) = \sqrt{\pi}$$

$$2^{2x-1}\Gamma(x)\Gamma\left(x + \frac{1}{2}\right) = \sqrt{\pi}\,\Gamma(2x)$$

3. Laplace Transforms

The Laplace transform of the function $f(t)$, denoted by $F(s)$ or $L\{f(t)\}$, is defined

$$F(s) = \int_0^\infty f(t)e^{-st}\,dt$$

provided that the integration may be validly performed. A sufficient condition for the existence of $F(s)$ is that $f(t)$ be of exponential order as $t \to \infty$ and that it is sectionally continuous over every finite interval in the range $t \geq 0$. The Laplace transform of $g(t)$ is denoted by $L\{g(t)\}$ or $G(s)$.

- *Operations*

$f(t)$	$F(s) = \int_0^\infty f(t)e^{-st}\,dt$
$af(t) + bg(t)$	$aF(s) + bG(s)$
$f'(t)$	$sF(s) - f(0)$
$f''(t)$	$s^2 F(s) - sf(0) - f'(0)$
$f^{(n)}(t)$	$s^n F(s) - s^{n-1} f(0)$ $- s^{n-2} f'(0)$ $- \cdots - f^{(n-1)}(0)$
$tf(t)$	$-F'(s)$
$t^n f(t)$	$(-1)^n F^{(n)}(s)$
$e^{at} f(t)$	$F(s - a)$

$$\int_0^t f(t-\beta) \cdot g(\beta)\, d\beta \qquad\qquad F(s) \cdot G(s)$$

$$f(t-a) \qquad\qquad e^{-as}F(s)$$

$$f\left(\frac{t}{a}\right) \qquad\qquad aF(as)$$

$$\int_0^t g(\beta)\, d\beta \qquad\qquad \frac{1}{s}G(s)$$

$$f(t-c)\delta(t-c) \qquad\qquad e^{-cs}F(s), \qquad c>0$$

where

$$\delta(t-c) = 0 \text{ if } 0 \le t < c$$
$$\qquad\quad = 1 \text{ if } t \ge c$$

$$f(t) = f(t+\omega) \qquad\qquad \frac{\displaystyle\int_0^\omega e^{-s\tau} f(\tau)\, d\tau}{1-e^{-s\omega}}$$
(periodic)

- *Table of Laplace Transforms*

$f(t)$	$F(s)$
1	$1/s$
t	$1/s^2$
$\dfrac{t^{n-1}}{(n-1)!}$	$1/s^n \qquad (n=1,2,3,\ldots)$
$\sqrt{t}$	$\dfrac{1}{2s}\sqrt{\dfrac{\pi}{s}}$

$\dfrac{1}{\sqrt{t}}$	$\sqrt{\dfrac{\pi}{s}}$
e^{at}	$\dfrac{1}{s-a}$
te^{at}	$\dfrac{1}{(s-a)^2}$
$\dfrac{t^{n-1}e^{at}}{(n-1)!}$	$\dfrac{1}{(s-a)^n}$ $\quad (n=1,2,3,\ldots)$
$\dfrac{t^x}{\Gamma(x+1)}$	$\dfrac{1}{s^{x+1}}$ $\quad (x>-1)$
$\sin at$	$\dfrac{a}{s^2+a^2}$
$\cos at$	$\dfrac{s}{s^2+a^2}$
$\sinh at$	$\dfrac{a}{s^2-a^2}$
$\cosh at$	$\dfrac{s}{s^2-a^2}$
$e^{at}-e^{bt}$	$\dfrac{a-b}{(s-a)(s-b)}$ $\quad (a\neq b)$
$ae^{at}-be^{bt}$	$\dfrac{s(a-b)}{(s-a)(s-b)}$ $\quad (a\neq b)$
$t\sin at$	$\dfrac{2as}{(s^2+a^2)^2}$
$t\cos at$	$\dfrac{s^2-a^2}{(s^2+a^2)^2}$

$e^{at} \sin bt$	$\dfrac{b}{(s-a)^2 + b^2}$
$e^{at} \cos bt$	$\dfrac{s-a}{(s-a)^2 + b^2}$
$\dfrac{\sin at}{t}$	$\text{Arctan}\, \dfrac{a}{s}$
$\dfrac{\sinh at}{t}$	$\dfrac{1}{2} \log_e \left(\dfrac{s+a}{s-a} \right)$

4. z-Transform

For the real-valued sequence $\{f(k)\}$ and complex vari-able z, the z-transform, $F(z) = Z\{f(k)\}$ is defined by

$$Z\{f(k)\} = F(z) = \sum_{k=0}^{\infty} f(k) z^{-k}$$

For example, the sequence $f(k) = 1$, $k = 0, 1, 2, \ldots$, has the z-transform

$$F(z) = 1 + z^{-1} + z^{-2} + z^{-3} \ldots + z^{-k} + \ldots.$$

• *z-Transform and the Laplace Transform*

For function $U(t)$ the output of the ideal sampler $U^*(t)$ is a set of values $U(kT)$, $k = 0, 1, 2, \ldots$, that is,

$$U^*(t) = \sum_{k=0}^{\infty} U(t)\, \delta(t - kT)$$

The Laplace transform of the output is

$$\mathscr{L}\{U^*(t)\} = \int_0^\infty e^{-st}U^*(t)\,dt = \int_0^\infty e^{-st} \sum_{k=0}^\infty U(t)\delta(t-kT)\,dt$$

$$= \sum_{k=0}^\infty e^{-skT}U(kT)$$

Defining $z = e^{sT}$ gives

$$\mathscr{L}\{U^*(t)\} = \sum_{k=0}^\infty U(kT)z^{-k}$$

which is the z-transform of the sampled signal $U(kT)$.

- *Properties*

    Linearity: $Z\{af_1(k) + bf_2(k)\} = aZ\{f_1(k)\} + bZ\{f_2(k)\}$
    $\qquad\qquad\qquad\quad = aF_1(z) + bF_2(z)$

    Right-shifting property: $Z\{f(k-n)\} = z^{-n}F(z)$

    Left-shifting property: $Z\{f(k+n)\} = z^n F(z)$
    $$- \sum_{k=0}^{n-1} f(k)z^{n-k}$$

    Time scaling: $Z\{a^k f(k)\} = F(z/a)$

Multiplication by k: $Z\{kf(k)\} = -z\,dF(z)/dz$

Initial value: $f(0) = \lim_{z \to \infty} (1 - z^{-1})F(z) = F(\infty)$

Final value: $\lim_{k \to \infty} f(k) = \lim_{z \to 1} (1 - z^{-1})F(z)$

Convolution: $Z\{f_1(k)*f_2(k)\} = F_1(z)F_2(z)$

- *z-Transforms of Sampled Functions*

$f(k)$	$Z\{f(kT)\} = F(z)$
1 at k; else 0	z^{-k}
1	$\dfrac{z}{z-1}$
kT	$\dfrac{Tz}{(z-1)^2}$
$(kT)^2$	$\dfrac{T^2 z(z+1)}{(z-1)^3}$
$\sin \omega kT$	$\dfrac{z \sin \omega T}{z^2 - 2z \cos \omega T + 1}$
$\cos \omega T$	$\dfrac{z(z - \cos \omega T)}{z^2 - 2z \cos \omega T + 1}$
e^{-akT}	$\dfrac{z}{z - e^{-aT}}$
kTe^{-akT}	$\dfrac{zTe^{-aT}}{(z - e^{-aT})^2}$

$(kT)^2 e^{-akT}$	$\dfrac{T^2 e^{-aT} z(z + e^{-aT})}{(z - e^{-aT})^3}$
$e^{-akT} \sin \omega kT$	$\dfrac{z e^{-aT} \sin \omega T}{z^2 - 2z e^{-aT} \cos \omega T + e^{-2aT}}$
$e^{-akT} \cos \omega kT$	$\dfrac{z(z - e^{-aT} \cos \omega T)}{z^2 - 2z e^{-aT} \cos \omega T + e^{-2aT}}$
$a^k \sin \omega kT$	$\dfrac{az \sin \omega T}{z^2 - 2az \cos \omega T + a^2}$
$a^k \cos \omega kT$	$\dfrac{z(z - a \cos \omega T)}{z^2 - 2az \cos \omega T + a^2}$

5. Fourier Series

The periodic function $f(t)$, with period 2π may be represented by the trigonometric series

$$a_0 + \sum_1^\infty (a_n \cos nt + b_n \sin nt)$$

where the coefficients are determined from

$$a_0 = \frac{1}{2\pi} \int_{-\pi}^{\pi} f(t)\, dt$$

$$a_n = \frac{1}{\pi} \int_{-\pi}^{\pi} f(t) \cos nt\, dt$$

$$b_n = \frac{1}{\pi} \int_{-\pi}^{\pi} f(t) \sin nt\, dt \qquad (n = 1, 2, 3, \ldots)$$

Such a trigonometric series is called the Fourier series corresponding to $f(t)$ and the coefficients are termed Fourier coefficients of $f(t)$. If the function is piecewise continuous in the interval $-\pi \leq t \leq \pi$, and has left- and right-hand derivatives at each point in that interval, then the series is convergent with sum $f(t)$ except at points t_i at which $f(t)$ is discontinuous. At such points of discontinuity, the sum of the series is the arithmetic mean of the right- and left-hand limits of $f(t)$ at t_i. The integrals in the formulas for the Fourier coefficients can have limits of integration that span a length of 2π, for example, 0 to 2π (because of the periodicity of the integrands).

6. Functions with Period Other Than 2π

If $f(t)$ has period P the Fourier series is

$$f(t) \sim a_0 + \sum_{1}^{\infty} \left(a_n \cos \frac{2\pi n}{P} t + b_n \sin \frac{2\pi n}{P} t \right),$$

where

$$a_0 = \frac{1}{P} \int_{-P/2}^{P/2} f(t)\, dt$$

$$a_n = \frac{2}{P} \int_{-P/2}^{P/2} f(t) \cos \frac{2\pi n}{P} t\, dt$$

$$b_n = \frac{2}{P} \int_{-P/2}^{P/2} f(t) \sin \frac{2\pi n}{P} t\, dt.$$

Again, the interval of integration in these formulas may be replaced by an interval of length P, for example, 0 to P.

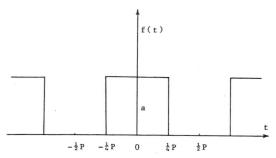

Figure 9.1. Square wave:

$$f(t) \sim \frac{a}{2} + \frac{2a}{\pi}\left(\cos\frac{2\pi t}{P} - \frac{1}{3}\cos\frac{6\pi t}{P} + \frac{1}{5}\cos\frac{10\pi t}{P} + \ldots\right).$$

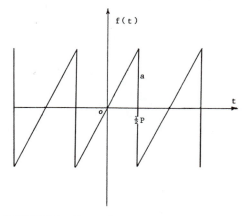

FIGURE 9.2. Sawtooth wave:

$$f(t) \sim \frac{2a}{\pi}\left(\sin\frac{2\pi t}{P} - \frac{1}{2}\sin\frac{4\pi t}{P} + \frac{1}{3}\sin\frac{6\pi t}{P} - \ldots\right).$$

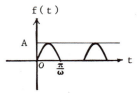

FIGURE 9.3. Half-wave rectifier:

$$f(t) \sim \frac{A}{\pi} + \frac{A}{2} \sin \omega t$$
$$- \frac{2A}{\pi} \left(\frac{1}{(1)(3)} \cos 2\omega t + \frac{1}{(3)(5)} \cos 4\omega t + \dots \right).$$

7. Bessel Functions

Bessel functions, also called cylindrical functions, arise in many physical problems as solutions of the differential equation

$$x^2 y'' + xy' + (x^2 - n^2)y = 0$$

which is known as Bessel's equation. Certain solutions of the above, known as *Bessel functions of the first kind of order n*, are given by

$$J_n(x) = \sum_{k=0}^{\infty} \frac{(-1)^k}{k!\Gamma(n+k+1)} \left(\frac{x}{2} \right)^{n+2k}$$

$$J_{-n}(x) = \sum_{k=0}^{\infty} \frac{(-1)^k}{k!\Gamma(-n+k+1)} \left(\frac{x}{2} \right)^{-n+2k}$$

98

In the above it is noteworthy that the gamma function must be defined for the negative argument q: $\Gamma(q) = \Gamma(q+1)/q$, provided that q is not a negative integer. When q is a negative integer, $1/\Gamma(q)$ is defined to be zero. The functions $J_{-n}(x)$ and $J_n(x)$ are solutions of Bessel's equation for all real n. It is seen, for $n = 1, 2, 3, \ldots$ that

$$J_{-n}(x) = (-1)^n J_n(x)$$

and, therefore, these are not independent; hence, a linear combination of these is not a general solution. When, however, n is not a positive integer, a negative integer, nor zero, the linear combination with arbitrary constants c_1 and c_2

$$y = c_1 J_n(x) + c_2 J_{-n}(x)$$

is the general solution of the Bessel differential equation.

The zero order function is especially important as it arises in the solution of the heat equation (for a "long" cylinder):

$$J_0(x) = 1 - \frac{x^2}{2^2} + \frac{x^4}{2^2 4^2} - \frac{x^6}{2^2 4^2 6^2} + \ldots$$

while the following relations show a connection to the trigonometric functions:

$$J_{\frac{1}{2}}(x) = \left[\frac{2}{\pi x} \right]^{1/2} \sin x$$

$$J_{-\frac{1}{2}}(x) = \left[\frac{2}{\pi x} \right]^{1/2} \cos x$$

The following recursion formula gives $J_{n+1}(x)$ for any order in terms of lower order functions:

$$\frac{2n}{x} J_n(x) = J_{n-1}(x) + J_{n+1}(x)$$

8. Legendre Polynomials

If Laplace's equation, $\nabla^2 V = 0$, is expressed in spherical coordinates, it is

$$r^2 \sin \theta \frac{\delta^2 V}{\delta r^2} + 2r \sin \theta \frac{\delta V}{\delta r} + \sin \theta \frac{\delta^2 V}{\delta \theta^2} + \cos \theta \frac{\delta V}{\delta \theta}$$

$$+ \frac{1}{\sin \theta} \frac{\delta^2 V}{\delta \phi^2} = 0$$

and any of its solutions, $V(r, \theta, \phi)$, are known as *spherical harmonics*. The solution as a product

$$V(r, \theta, \phi) = R(r)\Theta(\theta)$$

which is independent of ϕ, leads to

$$\sin^2 \theta \Theta'' + \sin \theta \cos \theta \Theta' + [n(n+1)\sin^2 \theta]\Theta = 0$$

Rearrangement and substitution of $x = \cos \theta$ leads to

$$(1 - x^2) \frac{d^2 \Theta}{dx^2} - 2x \frac{d\Theta}{dx} + n(n+1)\Theta = 0$$

known as *Legendre's equation*. Important special cases are those in which n is zero or a positive integer, and, for such cases, Legendre's equation is satisfied by poly-

nomials called Legendre polynomials, $P_n(x)$. A short
list of Legendre polynomials, expressed in terms of x
and $\cos \theta$, is given below. These are given by the
following general formula:

$$P_n(x) = \sum_{j=0}^{L} \frac{(-1)^j (2n-2j)!}{2^n j! (n-j)! (n-2j)!} x^{n-2j}$$

where $L = n/2$ if n is even and $L = (n-1)/2$ if n is
odd. Some are given below:

$P_0(x) = 1$

$P_1(x) = x$

$P_2(x) = \frac{1}{2}(3x^2 - 1)$

$P_3(x) = \frac{1}{2}(5x^3 - 3x)$

$P_4(x) = \frac{1}{8}(35x^4 - 30x^2 + 3)$

$P_5(x) = \frac{1}{8}(63x^5 - 70x^3 + 15x)$

$P_0(\cos \theta) = 1$

$P_1(\cos \theta) = \cos \theta$

$P_2(\cos \theta) = \frac{1}{4}(3 \cos 2\theta + 1)$

$P_3(\cos \theta) = \frac{1}{8}(5 \cos 3\theta + 3 \cos \theta)$

$$P_4(\cos\theta) = \frac{1}{64}(35\cos 4\theta + 20\cos 2\theta + 9)$$

Additional Legendre polynomials may be determined from the *recursion formula*

$$(n+1)P_{n+1}(x) - (2n+1)xP_n(x)$$

$$+ nP_{n-1}(x) = 0 \qquad (n = 1, 2, \dots)$$

or the *Rodrigues formula*

$$P_n(x) = \frac{1}{2^n n!}\frac{d^n}{dx^n}(x^2 - 1)^n$$

9. Laguerre Polynomials

Laguerre polynomials, denoted $L_n(x)$, are solutions of the differential equation

$$xy'' + (1 - x)y' + ny = 0$$

and are given by

$$L_n(x) = \sum_{j=0}^{n}\frac{(-1)^j}{j!}C_{(n,j)}x^j \qquad (n = 0, 1, 2, \dots)$$

Thus,

$$L_0(x) = 1$$

$$L_1(x) = 1 - x$$

$$L_2(x) = 1 - 2x + \frac{1}{2}x^2$$

$$L_3(x) = 1 - 3x + \frac{3}{2}x^2 - \frac{1}{6}x^3$$

Additional Laguerre polynomials may be obtained from the recursion formula

$$(n+1)L_{n+1}(x) - (2n+1-x)L_n(x)$$

$$+ nL_{n-1}(x) = 0$$

10. Hermite Polynomials

The Hermite polynomials, denoted $H_n(x)$, are given by

$$H_0 = 1, \quad H_n(x) = (-1)^n e^{x^2} \frac{d^n e^{-x^2}}{dx^n}$$

$$(n = 1, 2, \ldots)$$

and are solutions of the differential equation

$$y'' - 2xy' + 2ny = 0 \quad (n = 0, 1, 2, \ldots)$$

The first few Hermite polynomials are

$H_0 = 1$ $H_1(x) = 2x$
$H_2(x) = 4x^2 - 2$ $H_3(x) = 8x^3 - 12x$
$H_4(x) = 16x^4 - 48x^2 + 12$

Additional Hermite polynomials may be obtained from the relation

$$H_{n+1}(x) = 2xH_n(x) - H_n'(x),$$

where prime denotes differentiation with respect to x.

11. Orthogonality

A set of functions $\{f_n(x)\}$ $(n = 1, 2, \ldots)$ is orthogonal in an interval (a, b) with respect to a given weight function $w(x)$ if

$$\int_a^b w(x) f_m(x) f_n(x)\, dx = 0 \qquad \text{when } m \neq n$$

The following polynomials are orthogonal on the given interval for the given $w(x)$:

Legendre polynomials: $\quad P_n(x) \quad w(x) = 1$
$$a = -1, b = 1$$

Laguerre polynomials: $\quad L_n(x) \quad w(x) = \exp(-x)$
$$a = 0, b = \infty$$

Hermite polynomials: $\quad H_n(x) \quad w(x) = \exp(-x^2)$
$$a = -\infty, b = \infty$$

The Bessel functions *of order* n, $J_n(\lambda_1 x)$, $J_n(\lambda_2 x), \ldots$, are orthogonal with respect to $w(x) = x$ over the interval $(0, c)$ provided that the λ_i are the positive roots of $J_n(\lambda c) = 0$:

$$\int_0^c x J_n(\lambda_j x) J_n(\lambda_k x)\, dx = 0 \qquad (j \neq k)$$

where n is fixed and $n \geq 0$.

10 Differential Equations

1. First Order-First Degree Equations

$$M(x,y)\,dx + N(x,y)\,dy = 0$$

a. If the equation can be put in the form $A(x)\,dx + B(y)\,dy = 0$, it is *separable* and the solution follows by integration: $\int A(x)\,dx + \int B(y)\,dy = C$; thus, $x(1+y^2)\,dx + y\,dy = 0$ is separable since it is equivalent to $x\,dx + y\,dy/(1+y^2) = 0$, and integration yields $x^2/2 + \frac{1}{2}\log(1+y^2) + C = 0$.

b. If $M(x,y)$ and $N(x,y)$ are *homogeneous* and of the *same degree* in x and y, then substitution of vx for y (thus, $dy = v\,dx + x\,dv$) will yield a separable equation in the variables x and y. [A function such as $M(x,y)$ is homogeneous of degree n in x and y if $M(cx,cy) = c^n M(x,y)$.] For example, $(y-2x)\,dx + (2y+x)\,dy$ has M and N each homogeneous and of degree one so that substitution of $y = vx$ yields the separable equation

$$\frac{2}{x}\,dx + \frac{2v+1}{v^2+v-1}\,dv = 0.$$

c. If $M(x,y)\,dx + N(x,y)\,dy$ is the differential of some function $F(x,y)$, then the given equation is said to be *exact*. A necessary and sufficient condition for exactness is $\partial M / \partial y = \partial N / \partial x$. When the equation is exact, F is found from the relations $\partial F / \partial x = M$ and $\partial F / \partial y = N$, and the solution is $F(x, y) = C$ (constant). For example, $(x^2 + y)\, dy + (2xy - 3x^2)\, dx$ is exact since $\partial M / \partial y = 2x$ and $\partial N / \partial x = 2x$. F is found from $\partial F / \partial x = 2xy - 3x^2$ and $\partial F / \partial y = x^2 + y$. From the first of these, $F = x^2 y - x^3 + \phi(y)$; from the second, $F = x^2 y + y^2 / 2 + \Psi(x)$. It follows that $F = x^2 y - x^3 + y^2 / 2$, and $F = C$ is the solution.

d. Linear, order one in y: Such an equation has the form $dy + P(x) y\, dx = Q(x)\, dx$. Multiplication by $\exp[\int P(x)\, dx]$ yields

$$d\left[y \exp\left(\int P\, dx \right) \right] = Q(x) \exp\left(\int P\, dx \right) dx.$$

For example, $dy + (2/x)y\, dy = x^2\, dx$ is linear in y. $P(x) = 2/x$, so $\int P\, dx = 2 \ln x = \ln x^2$, and $\exp(\int P\, dx) = x^2$. Multiplication by x^2 yields $d(x^2 y) = x^4\, dx$, and integration gives the solution $x^2 y = x^5 / 5 + C$.

2. *Second Order Linear Equations (With Constant Coefficients)*

$$(b_0 D^2 + b_1 D + b_2) y = f(x), \qquad D = \frac{d}{dx}.$$

a. Right-hand side $= 0$ (homogeneous case)

$$(b_0 D^2 + b_1 D + b_2) y = 0.$$

The *auxiliary equation* associated with the above is

$$b_0 m^2 + b_1 m + b_2 = 0.$$

If the roots of the auxiliary equation are *real and distinct*, say m_1 and m_2, then the solution is

$$y = C_1 e^{m_1 x} + C_2 e^{m_2 x}$$

where the C's are arbitrary constants.

If the roots of the auxiliary equation are *real and repeated*, say $m_1 = m_2 = p$, then the solution is

$$y = C_1 e^{px} + C_2 x e^{px}.$$

If the roots of the auxiliary equation are *complex* $a + ib$ and $a - ib$, then the solution is

$$y = C_1 e^{ax} \cos bx + C_2 e^{ax} \sin bx.$$

b. Right-hand side $\neq 0$ (nonhomogeneous case)

$$(b_0 D^2 + b_1 D + b_2)y = f(x)$$

The general solution is $y = C_1 y_1(x) + C_2 y_2(x) + y_p(x)$ where y_1 and y_2 are solutions of the corresponding homogeneous equation and y_p is a solution of the given nonhomogeneous differential equation. y_p has the form $y_p(x) = A(x) y_1(x) + B(x) y_2(x)$ and A and B are found from simultaneous solution of $A' y_1 + B' y_2 = 0$ and $A' y_1' + B' y_2' = f(x)/b_0$. A solution exists if the determinant

$$\begin{vmatrix} y_1 & y_2 \\ y_1' & y_2' \end{vmatrix}$$

does not equal zero. The simultaneous equations yield A' and B' from which A and B follow by integration. For example,

$$(D^2 + D - 2)y = e^{-3x}.$$

The auxiliary equation has the distinct roots 1 and -2; hence $y_1 = e^x$ and $y_2 = e^{-2x}$, so that $y_p = Ae^x + Be^{-2x}$. The simultaneous equations are

$$A'e^x - 2B'e^{-2x} = e^{-3x}$$

$$A'e^x + B'e^{-2x} = 0$$

and give $A' = (1/3)e^{-4x}$ and $B' = (-1/3)e^{-x}$. Thus, $A = (-1/12)e^{-4x}$ and $B = (1/3)e^{-x}$ so that

$$y_p = (-1/12)e^{-3x} + (1/3)e^{-3x}$$

$$= \tfrac{1}{4}e^{-3x}.$$

$$\therefore y = C_1 e^x + C_2 e^{-2x} + \tfrac{1}{4}e^{-3x}.$$

11 Statistics

1. Arithmetic Mean

$$\mu = \frac{\Sigma X_i}{N},$$

where X_i is a measurement in the population and N is the total number of X_i in the population. For a *sample* of size n the sample mean, denoted $\overline{X}$, is

$$\overline{X} = \frac{\Sigma X_i}{n}.$$

2. Median

The median is the middle measurement when an odd number (n) of measurements is arranged in order; if n is even, it is the midpoint between the two middle measurements.

3. Mode

It is the most frequently occurring measurement in a set.

4. Geometric Mean

$$\text{geometric mean} = \sqrt[n]{X_1 X_2 \ldots X_n}$$

5. Harmonic Mean

The harmonic mean H of n numbers $X_1, X_2, \ldots, X_n$, is

$$H = \frac{n}{\Sigma(1/X_i)}$$

6. Variance

The mean of the sum of squares of deviations from the mean (μ) is the population variance, denoted σ^2

$$\sigma^2 = \Sigma(X_i - \mu)^2/N.$$

The sample variance, s^2, for sample size n is

$$s^2 = \Sigma(X_i - \overline{X})^2/(n-1).$$

A simpler computational form is

$$s^2 = \frac{\Sigma X_i^2 - \dfrac{(\Sigma X_i)^2}{n}}{n-1}$$

7. Standard Deviation

The positive square root of the population variance is the standard deviation. For a population

$$\sigma = \left[\frac{\Sigma X_i^2 - \dfrac{(\Sigma X_i)^2}{N}}{N} \right]^{1/2};$$

for a sample

$$s = \left[\frac{\sum X_i^2 - \dfrac{(\sum X_i)^2}{n}}{n-1} \right]^{1/2}.$$

8. Coefficient of Variation

$$V = s/\overline{X}.$$

9. Probability

For the sample space U, with subsets A of U (called "events"), we consider the probability measure of an event A to be a real-valued function p defined over all subsets of U such that:

$0 \le p(A) \le 1$
$p(U) = 1$ and $p(\Phi) = 0$
If A_1 and A_2 are subsets of U
$p(A_1 \cup A_2) = p(A_1) + p(A_2) - p(A_1 \cap A_2)$

Two events A_1 and A_2 are called mutually exclusive if and only if $A_1 \cap A_2 = \phi$ (null set). These events are said to be independent if and only if $p(A_1 \cap A_2) = p(A_1)p(A_2)$.

- *Conditional Probability and Bayes' Rule*

The probability of an event A, given that an event B has occurred, is called the conditional probability and is denoted $p(A/B)$. Further

$$p(A/B) = \frac{p(A \cap B)}{p(B)}$$

111

Bayes' rule permits a calculation of *a posteriori* probability from given *a priori* probabilities and is stated below:

If $A_1, A_2, \ldots, A_n$ are n mutually exclusive events, and $p(A_1) + p(A_2) + \ldots + p(A_n) = 1$, and B is any event such that $p(B)$ is not 0, then the conditional probability $p(A_i/B)$ for any one of the events A_i, *given that B has occurred* is

$$p(A_i/B) = \frac{p(A_i)p(B/A_i)}{p(A_1)p(B/A_1) + p(A_2)p(B/A_2) + \ldots + p(A_n)p(B/A_n)}$$

Example
Among 5 different laboratory tests for detecting a certain disease, one is effective with probability 0.75, whereas each of the others is effective with probability 0.40. A medical student, unfamiliar with the advantage of the best test, selects one of them and is successful in detecting the disease in a patient. What is the probability that the most effective test was used?

Let B denote (the event) of detecting the disease, A_1 the selection of the best test, and A_2 the selection of one of the other 4 tests; thus, $p(A_1) = 1/5$, $p(A_2) = 4/5$, $p(B/A_1) = 0.75$ and $p(B/A_2) = 0.40$. Therefore

$$p(A_1/B) = \frac{\frac{1}{5}(0.75)}{\frac{1}{5}(0.75) + \frac{4}{5}(0.40)} = 0.319$$

Note, the *a priori* probability is 0.20; the outcome raises this probability to 0.319.

10. Binomial Distribution

In an experiment consisting of n independent trials in which an event has probability p in a single trial, the probability P_X of obtaining X successes is given by

$$P_X = C_{(n,X)} p^X q^{(n-X)}$$

where

$$q = (1-p) \quad \text{and} \quad C_{(n,X)} = \frac{n!}{X!(n-X)!}.$$

The probability of between a and b successes (both a and b included) is $P_a + P_{a+1} + \ldots + P_b$, so if $a = 0$ and $b = n$, this sum is

$$\sum_{X=0}^{n} C_{(n,X)} p^X q^{(n-X)} = q^n + C_{(n,1)} q^{n-1} p$$

$$+ C_{(n,2)} q^{n-2} p^2 + \ldots + p^n = (q+p)^n = 1.$$

11. Mean of Binomially Distributed Variable

The mean number of successes in n independent trials is $m = np$ with standard deviation $\sigma = \sqrt{npq}$.

12. Normal Distribution

In the binomial distribution, as n increases the histogram of heights is approximated by the bell-shaped curve (normal curve)

$$Y = \frac{1}{\sigma\sqrt{2\pi}} e^{-(x-m)^2/2\sigma^2}$$

where $m =$ the mean of the binomial distribution $= np$, and $\sigma = \sqrt{npq}$ is the standard deviation. For any normally distributed random variable X with mean m and standard deviation σ the probability function (density) is given by the above.

The *standard* normal probability curve is given by

$$y = \frac{1}{\sqrt{2\pi}} e^{-Z^2/2}$$

and has mean $= 0$ and standard deviation $= 1$. The total area under the standard normal curve is 1. Any normal variable X can be put into standard form by defining $Z = (X - m)/\sigma$; thus the probability of X between a given X_1 and X_2 is the area under the standard normal curve between the corresponding Z_1 and Z_2. The standard normal curve is often used instead of the binomial distribution in experiments with discrete outcomes. For example, to determine the probability of obtaining 60 to 70 heads in a toss of 100 coins, we take $X = 59.5$ to $X = 70.5$ and compute corresponding values of Z from mean $np = 100 \frac{1}{2} = 50$, and the standard deviation $\sigma = \sqrt{(100)(1/2)(1/2)} = 5$. Thus, $Z = (59.5 - 50)/5 = 1.9$ and $Z = (70.5 - 50)/5 = 4.1$. The area between $Z = 0$ and $Z = 4.1$ is 0.5000 and between $Z = 0$ and $Z = 1.9$ is 0.4713; hence, the desired probability is 0.0287. The binomial distribution requires a more lengthy computation

$$C_{(100,60)}(1/2)^{60}(1/2)^{40} + C_{(100,61)}(1/2)^{61}(1/2)^{39}$$

$$+ \ldots + C_{(100,70)}(1/2)^{70}(1/2)^{30}.$$

Note that the normal curve is symmetric, whereas the histogram of the binomial distribution is symmetric only if $p = q = 1/2$. Accordingly, when p (hence q) differs appreciably from $1/2$, the difference between probabilities computed by each increases. It is usually recommended that the normal approximation not be used if p (or q) is so small that np (or nq) is less than 5.

13. Poisson Distribution

$$P = \frac{e^{-m}m^r}{r!}$$

is an approximation to the binomial probability for r successes in n trials when $m = np$ is small (< 5) and the normal curve is not recommended to approximate binomial probabilities. The variance σ^2 in the Poisson distribution is np, the same value as the mean. ***Example***: A school's expulsion rate is 5 students per 1000. If class size is 400, what is the probability that 3 or more will be expelled? Since $p = 0.005$ and $n = 400$, $m = np = 2$, and $r = 3$. We obtain for $m = 2$ and $r(=x) = 3$ the probability $p = 0.323$.

14. Least Squares Regression

A set of n values (X_i, Y_i) that display a linear trend is described by the linear equation $\hat{Y}_i = \alpha + \beta X_i$. Variables α and β are constants (population parameters) and are the intercept and slope, respectively. The rule for determining the line is one minimizing the sum of the squared deviations

$$\sum_{i=1}^{n} (Y_i - \hat{Y}_i)^2$$

and with this *criterion* the parameters α and β are best estimated from a and b calculated as

$$b = \frac{\sum X_i Y_i - \dfrac{(\sum X_i)(\sum Y_i)}{n}}{\sum X_i^2 - \dfrac{(\sum X_i)^2}{n}}$$

and

$$a = \overline{Y} - b\overline{X},$$

where $\overline{X}$ and $\overline{Y}$ are mean values, assuming that for any value of X the distribution of Y values is normal with variances that are equal for all X and the latter (X) are obtained with negligible error. The null hypothesis, $H_0: \beta = 0$, is tested with analysis of variance:

Source	SS	DF	MS
Total $(Y_i - \overline{Y})$	$\sum(Y_i - \overline{Y})^2$	$n-1$	
Regression $(\hat{Y}_i - \overline{Y})$	$\sum(\hat{Y}_i - \overline{Y})^2$	1	
Residual $(Y_i - \hat{Y}_i)$	$\sum(Y_i - \hat{Y}_i)^2$	$n-2$	$\dfrac{SS_{resid.}}{(n-2)} = S_{Y \cdot X}^2$

Computing forms for SS terms are

$$SS_{total} = \Sigma(Y_i - \bar{Y})^2 = \Sigma Y_i^2 - (\Sigma Y_i)^2/n$$

$$SS_{regr.} = \Sigma(\hat{Y}_i - \bar{Y})^2 = \frac{[\Sigma X_i Y_i - (\Sigma X_i)(\Sigma Y_i)/n]^2}{\Sigma X_i^2 - (\Sigma X_i)^2/n}$$

Example: Given points: $(0, 1), (2, 3), (4, 9), (5, 16)$. Analysis proceeds with the following calculations. $\Sigma X = 11$; $\Sigma Y = 29$; $\Sigma X^2 = 45$; $\Sigma XY = 122$; $\bar{X} = 2.75$; $\bar{Y} = 7.25$; $b = 2.86$; $\Sigma(X_i - \bar{X})^2 = 14.7 \therefore \hat{Y}_i = -0.615 + 2.86X$.

	SS	DF	MS	
Total	136.7	3		$F = \dfrac{121}{7.85} = 15.4$
				(significant)
Regr.	121	1	121	
Resid.	15.7	2	$7.85 = S_{Y \cdot X}^2$	$r^2 = 0.885$;
				$s_b = 0.73$

$F = MS_{regr.}/MS_{resid.}$ is calculated and compared with the critical value of F for the desired confidence level for degrees of freedom 1 and $n - 2$. The coefficient of determination, denoted r^2, is

$$r^2 = SS_{regr.}/SS_{total}$$

r is the *correlation coefficient*. The *standard error of estimate* is $\sqrt{s_{Y \cdot X}^2}$ and is used to calculate confidence intervals for α and β. For the confidence limits of β and α

$$b \pm t s_{Y \cdot X} \sqrt{\frac{1}{\Sigma(X_i - \overline{X})^2}} \quad , \quad a \pm t s_{Y \cdot X} \sqrt{\frac{1}{n} + \frac{\overline{X}^2}{\Sigma(X_i - \overline{X})^2}}$$

where t has $n - 2$ degrees of freedom.

The null hypothesis $H_0: \beta = 0$, can also be tested with the t statistic:

$$t = \frac{b}{s_b}$$

where s_b is the standard error of b

$$s_b = \frac{s_{Y \cdot X}}{\left[\Sigma(X_i - \overline{X})^2 \right]^{1/2}}$$

- *Standard Error of $\hat{Y}$*

An estimate of the mean value of Y for a given value of X, say X_0, is given by the regression equation

$$\hat{Y}_0 = a + bX_0.$$

The standard error of this predicted value is given by

$$S_{\hat{Y}_0} = S_{Y \cdot X} \left[\frac{1}{n} + \frac{(X_0 - \overline{X})^2}{\Sigma(X_i - \overline{X})^2} \right]^{\frac{1}{2}}$$

and is a minimum when $X_0 = \overline{X}$ and increases as X_0 moves away from $\overline{X}$ in either direction.

15. Summary of Probability Distributions

• Continuous Distributions

Distribution

Normal

$$y = \frac{1}{\sigma\sqrt{2\pi}} \exp[-(x-m)^2/2\sigma^2]$$

Mean $= m$

Variance $= \sigma^2$

Standard normal

$$y = \frac{1}{\sqrt{2\pi}} \exp(-z^2/2)$$

Mean $= 0$

Variance $= 1$

F-distribution

$$y = A \frac{F^{\frac{f_1-2}{2}}}{(f_2 + f_1 F)^{\frac{f_1+f_2}{2}}};$$

$$\text{where } A = \frac{\Gamma\left(\frac{f_1+f_2}{2}\right)}{\Gamma\left(\frac{f_1}{2}\right)\Gamma\left(\frac{f_2}{2}\right)} f_1^{\frac{f_1}{2}} f_2^{\frac{f_2}{2}}$$

$$\text{Mean} = \frac{f_2}{f_2 - 2}$$

$$\text{Variance} = \frac{2f_2^2(f_1 + f_2 - 2)}{f_1(f_2 - 2)^2(f_2 - 4)}$$

Chi-square

$$y = \frac{1}{2^{f/2}\Gamma(f/2)}\exp(-\frac{1}{2}x^2)(x^2)^{\frac{f-2}{2}}$$

$\text{Mean} = f$

$\text{Variance} = 2f$

Students t

$$y = A(1 + t^2/f)^{-(f+1)/2}; \text{ where } A = \frac{\Gamma(f/2 + 1/2)}{\sqrt{f\pi}\,\Gamma(f/2)}$$

$\text{Mean} = 0$

$$\text{Variance} = \frac{f}{f-2} \quad (\text{for } f > 2)$$

- *Discrete Distributions*

Binomial distribution

$$y = C_{(n,x)}\,p^x(1-p)^{n-x}$$

$\text{Mean} = np$

$\text{Variance} = np\,(1-p)$

Poisson distribution

$$y = \frac{e^{-m}m^x}{x!}$$

Mean $= m$

Variance $= m$

12 Table of Derivatives

In the following table, a and n are constants, e is the base of the natural logarithms, and u and v denote functions of x.

1. $\dfrac{d}{dx}(a) = 0$

2. $\dfrac{d}{dx}(x) = 1$

3. $\dfrac{d}{dx}(au) = a\dfrac{du}{dx}$

4. $\dfrac{d}{dx}(u+v) = \dfrac{du}{dx} + \dfrac{dv}{dx}$

5. $\dfrac{d}{dx}(uv) = u\dfrac{dv}{dx} + v\dfrac{du}{dx}$

6. $\dfrac{d}{dx}\left(\dfrac{u}{v}\right) = \dfrac{v\dfrac{du}{dx} - u\dfrac{dv}{dx}}{v^2}$

7. $\dfrac{d}{dx}(u^n) = nu^{n-1}\dfrac{du}{dx}$

8. $\dfrac{d}{dx}e^u = e^u\dfrac{du}{dx}$

9. $\dfrac{d}{dx}a^u = (\log_e a)a^u\dfrac{du}{dx}$

10. $\dfrac{d}{dx} \log_e u = \left(\dfrac{1}{u} \right) \dfrac{du}{dx}$

11. $\dfrac{d}{dx} \log_a u = (\log_a e) \left(\dfrac{1}{u} \right) \dfrac{du}{dx}$

12. $\dfrac{d}{dx} u^v = v u^{v-1} \dfrac{du}{dx} + u^v (\log_e u) \dfrac{dv}{dx}$

13. $\dfrac{d}{dx} \sin u = \cos u \dfrac{du}{dx}$

14. $\dfrac{d}{dx} \cos u = -\sin u \dfrac{du}{dx}$

15. $\dfrac{d}{dx} \tan u = \sec^2 u \dfrac{du}{dx}$

16. $\dfrac{d}{dx} \operatorname{ctn} u = -\csc^2 u \dfrac{du}{dx}$

17. $\dfrac{d}{dx} \sec u = \sec u \tan u \dfrac{du}{dx}$

18. $\dfrac{d}{dx} \csc u = -\csc u \operatorname{ctn} u \dfrac{du}{dx}$

19. $\dfrac{d}{dx} \sin^{-1} u = \dfrac{1}{\sqrt{1-u^2}} \dfrac{du}{dx}$ $\qquad (-\tfrac{1}{2}\pi \le \sin^{-1} u \le \tfrac{1}{2}\pi)$

20. $\dfrac{d}{dx} \cos^{-1} u = \dfrac{-1}{\sqrt{1-u^2}} \dfrac{du}{dx}$ $\qquad (0 \le \cos^{-1} u \le \pi)$

21. $\dfrac{d}{dx} \tan^{-1} u = \dfrac{1}{1+u^2} \dfrac{du}{dx}$

123

22. $\dfrac{d}{dx} \operatorname{ctn}^{-1} u = \dfrac{-1}{1+u^2} \dfrac{du}{dx}$

23. $\dfrac{d}{dx} \sec^{-1} u = \dfrac{1}{u\sqrt{u^2-1}} \dfrac{du}{dx},$

$$\left(-\pi \le \sec^{-1} u < -\tfrac{1}{2}\pi; 0 \le \sec^{-1} u < \tfrac{1}{2}\pi\right)$$

24. $\dfrac{d}{dx} \csc^{-1} u = \dfrac{-1}{u\sqrt{u^2-1}} \dfrac{du}{dx},$

$$\left(-\pi < \csc^{-1} u \le -\tfrac{1}{2}\pi; 0 < \csc^{-1} u \le \tfrac{1}{2}\pi\right)$$

25. $\dfrac{d}{dx} \sinh u = \cosh u \dfrac{du}{dx}$

26. $\dfrac{d}{dx} \cosh u = \sinh u \dfrac{du}{dx}$

27. $\dfrac{d}{dx} \tanh u = \operatorname{sech}^2 u \dfrac{du}{dx}$

28. $\dfrac{d}{dx} \operatorname{ctnh} u = -\operatorname{csch}^2 u \dfrac{du}{dx}$

29. $\dfrac{d}{dx} \operatorname{sech} u = -\operatorname{sech} u \tanh u \dfrac{du}{dx}$

30. $\dfrac{d}{dx} \operatorname{csch} u = -\operatorname{csch} u \operatorname{ctnh} u \dfrac{du}{dx}$

31. $\dfrac{d}{dx} \sinh^{-1} u = \dfrac{1}{\sqrt{u^2+1}} \dfrac{du}{dx}$

32. $\dfrac{d}{dx}\cosh^{-1} u = \dfrac{1}{\sqrt{u^2-1}}\dfrac{du}{dx}$

33. $\dfrac{d}{dx}\tanh^{-1} u = \dfrac{1}{1-u^2}\dfrac{du}{dx}$

34. $\dfrac{d}{dx}\operatorname{ctnh}^{-1} u = \dfrac{-1}{u^2-1}\dfrac{du}{dx}$

35. $\dfrac{d}{dx}\operatorname{sech}^{-1} u = \dfrac{-1}{u\sqrt{1-u^2}}\dfrac{du}{dx}$

36. $\dfrac{d}{dx}\operatorname{csch}^{-1} u = \dfrac{-1}{u\sqrt{u^2+1}}\dfrac{du}{dx}$

Additional Relations with Derivatives

$$\frac{d}{dt}\int_a^t f(x)\,dx = f(t)$$
$$\frac{d}{dt}\int_t^a f(x)\,dx = -f(t)$$

If $x=f(y)$, then

$$\frac{dy}{dx} = \frac{1}{\dfrac{dx}{dy}}$$

If $y=f(u)$ and $u=g(x)$, then

$$\frac{dy}{dx} = \frac{dy}{du}\cdot\frac{du}{dx} \qquad \text{(chain rule)}$$

If $x=f(t)$ and $y=g(t)$, then

$$\frac{dy}{dx} = \frac{g'(t)}{f'(t)},$$

and

$$\frac{d^2y}{dx^2} = \frac{f'(t)g''(t) - g'(t)f''(t)}{[f'(t)]^3}$$

(*Note*: exponent in denominator is 3.)

13 Table of Integrals

Indefinite Integrals
Definite Integrals

Table of Indefinite Integrals

Basic Forms (all logarithms are to base e)

1. $\int dx = x + C$

2. $\int x^n \, dx = \dfrac{x^{n+1}}{n+1} + C \qquad (n \neq -1)$

3. $\int \dfrac{dx}{x} = \log x + C$

4. $\int e^x \, dx = e^x + C$

5. $\int a^x \, dx = \dfrac{a^x}{\log a} + C$

6. $\int \sin x \, dx = -\cos x + C$

7. $\int \cos x\, dx = \sin x + C$

8. $\int \tan x\, dx = -\log \cos x + C$

9. $\int \sec^2 x\, dx = \tan x + C$

10. $\int \csc^2 x\, dx = -\operatorname{ctn} x + C$

11. $\int \sec x \tan x\, dx = \sec x + C$

12. $\int \sin^2 x\, dx = \frac{1}{2}x - \frac{1}{2}\sin x \cos x + C$

13. $\int \cos^2 x\, dx = \frac{1}{2}x + \frac{1}{2}\sin x \cos x + C$

Form ax + b

14. $\int \log x \, dx = x \log x - x + C$

In the following list, a constant of integration C should be added to the result of each integration.

Form $ax + b$

15. $\int (ax+b)^m \, dx = \dfrac{(ax+b)^{m+1}}{a(m+1)}$

$(m \neq -1)$

16. $\int x(ax+b)^m \, dx = \dfrac{(ax+b)^{m+2}}{a^2(m+2)} - \dfrac{b(ax+b)^{m+1}}{a^2(m+1)},$

$(m \neq -1, -2)$

17. $\int \dfrac{dx}{ax+b} = \dfrac{1}{a} \log(ax+b)$

Trigonometric Forms

18. $\displaystyle \int \frac{dx}{(ax+b)^2} = -\frac{1}{a(ax+b)}$

19. $\displaystyle \int (\sin ax)\, dx = -\frac{1}{a}\cos ax$

20. $\displaystyle \int (\sin^2 ax)\, dx = -\frac{1}{2a}\cos ax \sin ax + \frac{1}{2}x = \frac{1}{2}x - \frac{1}{4a}\sin 2ax$

21. $\displaystyle \int \sin(a+bx)\, dx = -\frac{1}{b}\cos(a+bx)$

22. $\displaystyle \int \frac{dx}{1 \pm \sin ax} = \mp \frac{1}{a}\tan\left(\frac{\pi}{4} \mp \frac{ax}{2}\right)$

23. $\displaystyle \int \frac{\sin ax}{1 \pm \sin ax}\, dx = \pm x + \frac{1}{a}\tan\left(\frac{\pi}{4} \mp \frac{ax}{2}\right)$

24. $\displaystyle \int (\cos ax)\, dx = \frac{1}{a}\sin ax$

Form ax + b and Trigonometric Forms

Trigonometric, Logarithmic, and Exponential Forms

25. $\int (\cos^2 ax)\,dx = \frac{1}{2a}\sin ax \cos ax + \frac{1}{2}x = \frac{1}{2}x + \frac{1}{4a}\sin 2ax$ $(m^2 \neq n^2)$

26. $\int (\sin mx)(\sin nx)\,dx = \dfrac{\sin(m-n)x}{2(m-n)} - \dfrac{\sin(m+n)x}{2(m+n)}$ $(m^2 \neq n^2)$

27. $\int (\cos mx)(\cos nx)\,dx = \dfrac{\sin(m-n)x}{2(m-n)} + \dfrac{\sin(m+n)x}{2(m+n)}$ $(m^2 \neq n^2)$

28. $\int (\sin ax)(\cos ax)\,dx = \dfrac{1}{2a}\sin^2 ax$

29. $\int (\sin mx)(\cos nx)\,dx = -\dfrac{\cos(m-n)x}{2(m-n)} - \dfrac{\cos(m+n)x}{2(m+n)}$ $(m^2 \neq n^2)$

Logarithmic Forms

30. $\int (\log x)\,dx = x\log x - x$

31. $\int x(\log x)\,dx = \dfrac{x^2}{2}\log x - \dfrac{x^2}{4}$

32. $\int x^2(\log x)\,dx = \dfrac{x^3}{3}\log x - \dfrac{x^3}{9}$

Exponential Forms

33. $\int e^x\,dx = e^x$

34. $\int e^{-x}\,dx = -e^{-x}$

35. $\int e^{ax}\,dx = \dfrac{e^{ax}}{a}$

36. $\int xe^{ax}\,dx = \dfrac{e^{ax}}{a^2}(ax - 1)$

37. $\int \dfrac{dx}{1 + e^x} = x - \log(1 + e^x) = \log\dfrac{e^x}{1 + e^x}$

Trigonometric, Logarithmic, and Exponential Forms

Table of Definite Integrals

Table of Definite Integrals

38. $\int_1^\infty \dfrac{dx}{x^m} = \dfrac{1}{m-1}$ $(m > 1)$

39. $\int_0^\infty \dfrac{dx}{(1+x)\sqrt{x}} = \pi$

40. $\int_0^\infty \dfrac{a\,dx}{a^2+x^2} = \dfrac{\pi}{2}$, if $a > 0$; 0, if $a = 0$; $-\dfrac{\pi}{2}$, if $a < 0$

41. $\int_0^\infty e^{-ax}\,dx = \dfrac{1}{a}$ $(a > 0)$

42. $\int_0^\infty \dfrac{e^{-ax} - e^{-bx}}{x}\,dx = \log\dfrac{b}{a}$ $(a, b > 0)$

14 Resistor Circuits

1. Electric Current and Voltage

We can express current as

$$i = \frac{dq}{dt}$$

The unit of current is the ampere (A); an ampere is 1 coulomb per second.

Current is the time rate of flow of electric charge. *Charge* is the quantity of electricity responsible for electric phenomena.

$$q = \int_0^t i\,d\tau + q(0)$$

The *voltage* across an element is the work required to move a positive charge of 1 coulomb from the first terminal through the element to the second terminal (the unit of voltage is the volt, V):

$$v = \frac{dw}{dq}$$

where v is voltage, w is energy, and q is charge. A charge of 1 coulomb delivers an energy of 1 joule as it moves through a voltage of 1 volt.

Power is the time rate of expending or absorbing energy. Thus, we have the equation

$$p = \frac{dw}{dt}$$

where p is the power in watts, w is energy in joules, and t is the time in seconds;

$$p = v \cdot i$$

2. Current Flow in a Circuit Element

When energy is delivered to the element, the voltage drop across two terminals a–b is said to be a voltage v as shown in Figure 14.1.

A *passive element* absorbs energy,

$$w = \int_{-\infty}^{t} v \, i \, d\tau \geq 0$$

when both v and i are the same sign.

3. Resistance and Ohm's Law

Resistance is the physical property of an element or device that impedes the flow of current; it is represented by the symbol R. Resistance R is defined as

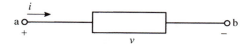

FIGURE 14.1.

$$R = \frac{\rho L}{A}$$

where A is the cross-sectional area, $\dot{\rho}$ is the resistivity, L is the length, and v is the voltage across the wire element.

Ohm's law, which relates the voltage and current of a resistance, is

$$v = Ri$$

The unit of resistance R was named the ohm in honor of Ohm and is usually abbreviated by the symbol Ω (capital omega), where $1\ \Omega = 1$ V/A.

Ohm's law can also be written as

$$i = Gv$$

where G denotes the conductance in siemens (S).

The power delivered to a resistor is

$$p = vi = \frac{v^2}{R} = i^2 R$$

4. Kirchhoff's Laws

By *Kirchhoff's current law* (KCL), the algebraic sum of the currents into a node at any instant is zero:

$$\sum_{n=1}^{N} i_n = 0$$

By *Kirchhoff's voltage law* (KVL), the algebraic sum of the voltages around any closed path in a circuit is identically zero for all time:

$$\sum_{n=1}^{N} v_n = 0$$

5. Voltage and Current Divider Circuits

The voltage, v_n, across the nth resistor of N resistors connected in series is

$$v_n = \frac{R_n}{R_1 + R_2 + \ldots + R_N} v_s = \frac{R_n}{\sum_{j=1}^{N} R_j}$$

where v_s is the source voltage connected in series with the resistors.

The current, i_n, in the conductance G_n connected in a parallel set of N conductances is

$$i_n = \frac{G_n i_s}{\sum_{j=1}^{N} G_j}$$

where i_s is a source current connected in parallel with the parallel set of conductances.

6. Equivalent Resistance and Equivalent Conductance

An equivalent resistance, R_s, for a series connection of N resistors is

$$R_s = \sum_{j=1}^{N} R_j$$

An equivalent conductance, G_p, for a parallel connection of N conductances is

$$G_p = \sum_{j=1}^{N} G_j$$

7. Node Voltages

The node voltage matrix equation for a circuit with N unknown node voltages is

$$\mathbf{G}\mathbf{v} = \mathbf{i}_s$$

where

$$\mathbf{v} = \begin{bmatrix} v_a \\ v_b \\ \vdots \\ v_N \end{bmatrix}$$

which is the vector of unknown node voltages. The matrix

$$\mathbf{i}_s = \begin{bmatrix} i_{s1} \\ i_{s2} \\ \vdots \\ i_{sN} \end{bmatrix}$$

is the vector consisting of the N current sources where i_{sn} is the sum of all the source currents into the node n.

139

When there are no dependent sources within the circuit, the conductance matrix is symmetric as

$$\mathbf{G} = \begin{bmatrix} \sum_a G & -G_{ab} & \cdots & -G_{aN} \\ -G_{ab} & \sum_b G & \cdots & -G_{bN} \\ \vdots & & & \\ -G_{aN} & -G_{bN} & \cdots & \sum_N G \end{bmatrix}$$

where $\sum_n G$ is the sum of the conductances at node n and G_{ij} is the sum of the conductances connecting node i and j. When the circuit includes dependent sources, the **G** matrix is not symmetric.

8. Mesh Current Analysis

We assume a planar network with N meshes containing N mesh currents flowing clockwise. The matrix equation for mesh current analysis with no dependent sources is

$$\mathbf{Ri} = \mathbf{v}_s$$

where **R** is a symmetric matrix with a diagonal consisting of the sum of resistances in each mesh, and the off-diagonal elements are the negative of the resistances connecting two meshes. The matrix **i** consists of the mesh currents as

$$\mathbf{i} = \begin{bmatrix} i_1 \\ i_2 \\ \vdots \\ i_N \end{bmatrix}$$

For N mesh currents. The source matrix $\mathbf{v}_s$ is

$$\mathbf{v}_s = \begin{bmatrix} v_{s1} \\ v_{s2} \\ \vdots \\ v_{sN} \end{bmatrix}$$

where v_{sj} is the sum of the sources in the jth mesh with the appropriate sign assigned to each source.

When dependent sources are present within the circuit the **R** matrix is not symmetric.

9. Source Transformations

A *source transformation* is a procedure for transforming one source into another while retaining the terminal characteristics of the original source. The transformation of a voltage source in series with a resistance R_s into a current source in parallel with a resistance R_p is summarized in Figure 14.2(a).

The transformation of a current source in parallel with a resistance R_p into a voltage source in series with a resistance R_s is summarized in Figure 14.2(b).

10. The Superposition Principle

The *superposition principle* may be stated as follows: In a linear circuit containing independent sources, the voltage across (or the current through) any element may be obtained by adding algebraically all the individual voltages (or currents) caused by each independent source acting alone, with all other independent voltage sources replaced by short circuits and all other independent current sources replaced by open circuits.

(a)

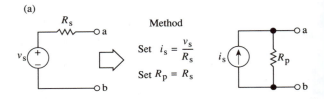

Method

Set $i_s = \dfrac{v_s}{R_s}$

Set $R_p = R_s$

(b)

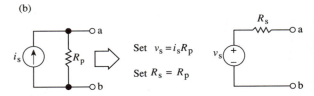

Set $v_s = i_s R_p$

Set $R_s = R_p$

FIGURE 14.2. Method of source transformations.

The voltage across an element, v, is

$$v = \sum_{j=1}^{N} v_j$$

where v_j is the voltage due to the jth source with all other sources disabled.

11. Thévenin's Theorem

Thévenin's theorem requires that, for any circuit of resistance elements and energy sources with an identified terminal pair, the circuit can be replaced by a series combination of an ideal voltage source v_t and a resistance R_t, where v_t is the open-circuit voltage at the two terminals and R_t is the ratio of the open-circuit voltage to the short-circuit current at the terminal pair:

$$v_t = v_{oc}$$

$$R_t = \frac{v_{oc}}{i_{sc}}$$

12. Norton's Theorem

Norton's theorem requires that, for any circuit of resistance elements and energy sources with an identified terminal pair, the circuit can be replaced by a parallel combination of an ideal current source i_n and a conductance G_n, where i_n is the short-circuit current at the two terminals and G_n is the ratio of the short-circuit current to the open-circuit voltage at the terminal pair:

$$i_n = i_{sc}$$

$$G_n = \frac{i_{sc}}{v_{oc}}$$

13. Tellegan's Theorem

Tellegan's theorem states that in an arbitrarily lumped network subject to KVL and KCL constraints, with reference directions of the branch currents and branch voltages associated with the KVL and KCL constraints, the product of all branch currents and branch voltages must equal zero. Tellegen's theorem may be summarized by the equation

$$\sum_{k=1}^{b} v_k j_k = 0$$

where the lower case letters v and j represent instantaneous values of the branch voltages and branch currents, respectively, and where b is the total number of branches. A matrix representation employing the branch current and branch voltage vectors also exists. Because $\mathbf{V}$ and $\mathbf{J}$ are column vectors we have

$$\mathbf{V} \cdot \mathbf{J} = \mathbf{V}^T \mathbf{J} = \mathbf{J}^T \mathbf{V}$$

14. Maximum Power Transfer

The *maximum power transfer* theorem states that the maximum power delivered by a source represented by its Thévenin equivalent circuit is attained when the load R_L is equal to the Thévenin resistance R_t (see Figure 14.3):

$$R_t = R_L \quad \text{for maximum power}$$

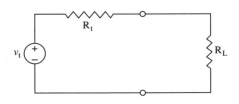

FIGURE 14.3.

144

15. Efficiency of Power Transfer

The *efficiency of power transfer* is defined as the ratio of the power delivered to the load, p_{out}, to the power supplied by the source, p_{in}, η as

$$\eta = p_{out}/p_{in}$$

15 Circuits with Energy Storage Elements

1. Capacitors

Capacitance is a measure of the ability of a device to store energy in the form of separated charge or in the form of an electric field:

$$q = Cv$$

where q is the charge, v is the voltage across the element, and C is the capacitance measured in farads (F).

The current through a capacitor is

$$i = C\frac{dv}{dt}$$

The voltage across a capacitor C is

$$v = \frac{1}{C}\int_{t_0}^{t} i\,d\tau + v(t_0)$$

where $v(t_0)$ is the voltage at t_0.

2. Inductors

Inductance is a measure of the ability of a device to store energy in the form of a magnetic field. The voltage across an inductor is

$$v = L \frac{di}{dt}$$

where i is the current through the inductor and L is the inductance measured in henrys (H).

The current in an inductor is

$$i = \frac{1}{L} \int_{t_0}^{t} v \, d\tau + i(t_0)$$

3. Energy Stored in Inductors and Capacitors

$$\omega = \frac{1}{2} C v^2$$

and

$$\omega = \frac{1}{2} L i^2$$

4. Series and Parallel Inductors

A series connection of N inductors can be represented by one series equivalent inductor L_s as

$$L_s = \sum_{n=1}^{N} L_n$$

A parallel connection of N inductors can be represented by one equivalent inductor L_p as

$$\frac{1}{L_p} = \sum_{n=1}^{N} \frac{1}{L_n}$$

147

5. Series and Parallel Capacitors

The equivalent capacitance of a set of N parallel capacitors is simply the sum of the individual capacitances:

$$C_p = \sum_{n=1}^{N} C_n$$

A series connection of N capacitors can be represented by one equivalent capacitance C_s:

$$\frac{1}{C_s} = \sum_{n=1}^{N} \frac{1}{C_n}$$

6. The Natural Response of an RL or RC Circuit

The *natural response* of a circuit depends only on the internal energy storage of the circuit and not on external sources. The natural response of a series connection of a resistor R and a capacitor C is

$$v = V_0 e^{-t/RC}$$

where $v(0) = V_0$ is the initial voltage on the capacitor and v is the capacitor voltage.

The natural response of a series connection of a resistor R and inductor L is

$$i = I_0 e^{-Rt/L}$$

where $i(0) = I_0$ is the initial current and i is the inductor current.

7. The Forced Response of an RL or RC Circuit Excited by a Constant Source

The *forced response* of a circuit is the behavior exhibited in reaction to one or more independent signal source. The forced response of an RC circuit is

$$v(t) = v(\infty) + [v(0) - v(\infty)]e^{-t/RC}$$

where $v(\infty)$ is the steady-state value at $t = \infty$.

The forced response of an RL circuit is

$$i(t) = i(\infty) + [i(0 - i(\infty)]e^{-t/\tau}$$

where $\tau = L/R$.

8. The Natural Response of a RLC Circuit

The differential equation for a parallel connection of an R, L, and C is

$$\frac{d^2v}{dt^2} + \frac{1}{RC}\frac{dv}{dt} + \frac{v}{LC} = 0$$

where v is the capacitor (see Figure 15.1).

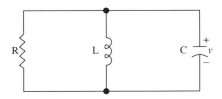

FIGURE 15.1. Parallel RLC circuit.

The differential equation for the series connection of R, L, and C is

$$\frac{d^2i}{dt^2} + \frac{R}{L}\frac{di}{dt} + \frac{i}{LC} = 0$$

where i is the current through the inductor (see Figure 15.2).

The characteristic equation is

$$s^2 + a_1 s + a_0 = 0$$

or

$$s^2 + 2\alpha s + \omega_0^2 = 0$$

Then the roots of the characteristic equation are

$$s_1 = -\alpha + \sqrt{\alpha^2 - \omega_0^2}$$

$$s_2 = -\alpha - \sqrt{\alpha^2 - \omega_0^2}$$

where $\omega_0 = 1/\sqrt{LC}$ is called the resonant frequency.

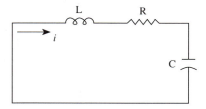

FIGURE 15.2. Series RLC circuit.

The roots of the characteristic equation assume three possible conditions.

1. Two real and distinct roots when $\alpha^2 > \omega_0^2$.

2. Two real equal roots when $\alpha^2 = \omega_0^2$.

3. Two complex roots when $\alpha^2 < \omega_0^2$.

When the two roots are real and distinct, the circuit is said to be *overdamped*. When the roots are both real and equal, the circuit is *critically damped*. When the two roots are complex conjugates, the circuit is said to be *underdamped*.

The overdamped natural response is

$$x = A_1 e^{-s_1 t} + A_2 e^{-s_2 t}$$

where $x = v$ for the parallel RLC circuit and $x = i$ for the series RLC circuit.

When the two roots are equal, the natural response is

$$x = e^{-\alpha t}(A_1 t + A_2)$$

When the circuit is underdamped, we have

$$x = e^{-\alpha t}(B_1 \cos \omega_d t + B_2 \sin \omega_d t)$$

where $\omega_d = \sqrt{\omega_0^2 - \alpha^2}$, the damped resonant frequency.

16 AC Circuits

1. Phasor Voltage and Current

A sinusoidal voltage or current at a given frequency is characterized by its amplitude and phase angle.

The current

$$i(t) = I_m \cos(\omega t + \theta)$$

is represented by the phasor

$$\mathbf{I} = I_m \underline{/\theta}$$

TABLE 16.1 Time Domain and Phasor Relationships for R, L, and C

Element	Time Domain	Frequency Domain
Resistor	$v = Ri$	$\mathbf{V} = R\mathbf{I}$
Inductor	$v = L\dfrac{di}{dt}$	$\mathbf{V} = j\omega L\mathbf{I}$
Capacitor	$i = C\dfrac{dv}{dt}$	$\mathbf{V} = \dfrac{1}{j\omega C}\mathbf{I}$

TABLE 16.2   Impedances of R, L, and C

Element	Impedance
Resistor	$\mathbf{Z} = R$
Inductor	$\mathbf{Z} = j\omega L$
Capacitor	$\mathbf{Z} = \dfrac{1}{j\omega C}$

2. Kirchhoff's Laws in the Phasor Form

Kirchhoff's current law requires the sum of the currents entering a node to be equal to zero:

$$\sum_{n=1}^{N} \mathbf{I}_n = 0$$

Kirchhoff's voltage law requires the sum of the voltages in a closed path to be zero:

$$\sum_{j=1}^{J} \mathbf{V}_j = 0$$

3. AC Steady-State Power

The instantaneous power delivered to a circuit element is

$$p(t) = v(t)i(t)$$

The average power delivered to an impedance $\mathbf{Z} = Z\underline{/\theta}$ is

$$P = \frac{V_m I_m}{2} \cos \theta$$

when $v = V_m \cos \omega t$ across the impedance.

4. *Maximum Power Transfer*

Maximum power is delivered to a load $\mathbf{Z}_L$ when $\mathbf{Z}_L$ is set equal to the complex conjugate of $\mathbf{Z}_t$, the Thévinin equivalent impedance of the circuit connected to the load:

$$\mathbf{Z}_L = \mathbf{Z}_t^*$$

5. *Effective Value of a Sinusoidal Waveform*

The effective value (rms value) of a sinusoidal voltage $v = V_m \cos \omega t$ is

$$V_{\text{eff}} = V_{\text{rms}} = \frac{V_m}{\sqrt{2}}$$

6. *Power Delivered to an Impedance Z*

$$P = V_{\text{eff}} I_{\text{eff}} \cos \theta$$

$$= VI \cos \theta \quad (W)$$

The power factor is

$$\text{pf} = \cos \theta$$

The apparent power is VI with units of voltamperes (VA). The reactive power is

$$Q = VI \sin \theta$$

with units of voltamperes reactive (VAR).

The complex power is

$$\mathbf{S} = P + jQ$$

$$= \mathbf{VI}^*$$

7. *Three-Phase Power*

A three-phase generator consists of three voltages

$$v_{a'a} = \sqrt{2}\,V \cos \omega t$$

$$v_{b'b} = \sqrt{2}\,V \cos(\omega t - 120°)$$

$$v_{c'c} = \sqrt{2}\,V \cos(\omega t - 240°)$$

or in phasor notation

$$\mathbf{V}_{a'a} = V\underline{/0°}$$

$$\mathbf{V}_{b'b} = V\underline{/-120°}$$

$$\mathbf{V}_{c'c} = V\underline{/-240°}$$

Because the voltage in phase $a'a$ reaches its maximum first, followed by that in phase $b'b$, and then by that in phase $c'c$, we say the *phase rotation* is *abc*.

A balanced load has equal load on each phase. A balanced three-phase system consists of three equal single-phase sources connected in Δ or Y supplying three equal loads connected in Δ or Y. For balanced three-phase systems in

$$\Delta: \quad \mathbf{V}_{line} = \mathbf{V}_{phase} \quad \text{and} \quad \mathbf{I}_{line} = \sqrt{3}\,\mathbf{I}_{phase} \underline{/-30°}$$

$$Y: \quad \mathbf{I}_{line} = \mathbf{I}_{phase} \quad \text{and} \quad \mathbf{V}_{line} = \sqrt{3}\,\mathbf{V}_{phase} \underline{/+30°}$$

In a Δ connection, the line current is $\sqrt{3}$ times the phase current and is displaced $-30°$ in phase; the line-to-line voltage is just equal to the phase voltage:

$$\mathbf{I}_{line} = \sqrt{3}\,_{phase} \underline{/-30°}$$

In a Y connection, the line-to-line voltage is $\sqrt{3}$ times the phase voltage and is displaced $30°$ in phase; the line current is just equal to the phase current:

$$\mathbf{V}_{line} = \sqrt{3}\,\mathbf{V}_{phase} \underline{/30°}$$

8. Power Calculations

The total power in a balanced three-phase load is the sum of three equal phase powers or

$$P_{total} = 3P_p = 3V_p I_p \cos\theta$$

where $\cos\theta$ is the power factor of the load.

For a Δ load,

$$P_{total} = \sqrt{3}\,\mathbf{V}_{line}\mathbf{I}_{line}\, cos\,\theta$$

9. The Reciprocity Theorem

In any passive, linear network, if a voltage **V** applied in branch 1 causes a current **I** to flow in branch 2, then voltage **V** applied in branch 2 will cause current **I** to flow in branch 1.

10. Model of the Transformer

The differential equations for the transformer model shown in Fig. 16.1 are

$$v_1 = L_1 \frac{di_1}{dt} + M \frac{di_2}{dt}$$

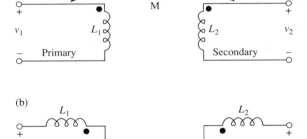

FIGURE 16.1. (a) Circuit symbol for the transformer. (b) Model of the transformer.

and

$$v_2 = L_2 \frac{di_2}{dt} + M \frac{di_1}{dt}$$

The phasor form of the transformer equations are

$$\mathbf{V}_1 = j\omega L_1 \mathbf{I}_1 + j\omega M \mathbf{I}_2$$

$$\mathbf{V}_2 = j\omega L_2 \mathbf{I}_2 + j\omega M \mathbf{I}_1$$

11. The Ideal Transformer

The model for the ideal transformer is shown in Figure 16.2.

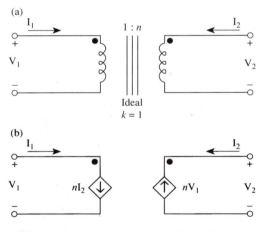

FIGURE 16.2. (a) Symbol for the ideal transformer. (b) Model for the ideal transformer.

17 T and Π and Two-Port Networks

1. T and Π Networks

A T network is shown in Figure 17.1(a) and a Π network is shown in Figure 17.1(b).

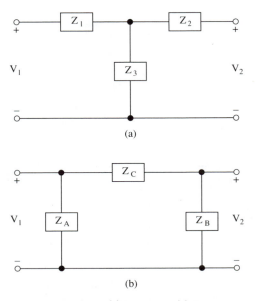

(a)

(b)

FIGURE 17.1. (a) T network. (b) Π network.

If a network has mirror-image symmetry with respect to some centerline, that is, if a line can be found to divide the network into two symmetrical halves, the network is a *symmetrical network*. The T network is symmetrical when $Z_1 = Z_2$, and the Π network is symmetrical when $Z_A = Z_B$. Furthermore, if all the impedances in either the T or Π are equal, the T or Π is completely symmetrical.

To convert a Π to T network, relationships for Z_1, Z_2, and Z_3 must be obtained in terms of the impedance Z_A, Z_B, and Z_C. Then we have

$$Z_1 = \frac{Z_A Z_C}{Z_A + Z_B + Z_C}$$

$$Z_2 = \frac{Z_B Z_C}{Z_A + Z_B + Z_C}$$

$$Z_3 = \frac{Z_A Z_B}{Z_A + Z_B + Z_C}$$

To convert a T to a Π network we use the relationships

$$Z_A = \frac{Z_1 Z_2 + Z_2 Z_3 + Z_3 Z_1}{Z_2}$$

$$Z_B = \frac{Z_1 Z_2 + Z_2 Z_3 + Z_3 Z_1}{Z_1}$$

$$Z_C = \frac{Z_1 Z_2 + Z_2 Z_3 + Z_3 Z_1}{Z_3}$$

When a T or Π is completely symmetrical, the conversion equations reduce to

$$Z_T = \frac{Z_\pi}{3}$$

and

$$Z_\Pi = 3Z_T$$

where Z_T is the impedance in each leg of the T network and Z_Π is the impedance in each leg of the Π network.

2. Two-Port Networks

A two-port network is a circuit with two pairs of terminals (ports) at which excitation can be applied or response measured. (One terminal may be common to input and output). The general two-port is shown in Figure 17.2. The impedance parameters of a two-port network are expressed as

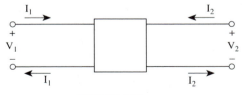

FIGURE 17.2.

$$V_1 = Z_{11}I_1 + Z_{12}I_2$$

$$V_2 = Z_{21}I_1 + Z_{22}I_2$$

The admittance parameters are expressed as

$$I_1 = Y_{11}V_1 + Y_{12}V_2$$

$$I_2 = Y_{21}V_1 + Y_{22}V_2$$

The hybrid h-parameters expressed in equation form are

$$V_1 = h_{11}I_1 + h_{12}V_2$$

$$I_2 = h_{21}I_1 + h_{22}V_2$$

The inverse hybrid parameter equations are

$$I_1 = g_{11}V_1 + g_{12}I_2$$

$$V_2 = g_{21}V_1 + g_{22}I_2$$

The transmission parameters are written as

$$V_1 = AV_2 - BI_2$$

$$I_1 = CV_2 - DI_2$$

The relationships between two-port parameters are summarized in the following Table.

Parameter Relationships

	Z		Y		h		g		T	
Z	Z_{11}	Z_{12}	$\dfrac{Y_{22}}{\Delta Y}$	$\dfrac{-Y_{12}}{\Delta Y}$	$\dfrac{\Delta h}{h_{22}}$	$\dfrac{h_{12}}{h_{22}}$	$\dfrac{1}{g_{11}}$	$\dfrac{-g_{12}}{g_{11}}$	$\dfrac{A}{C}$	$\dfrac{\Delta T}{C}$
	Z_{21}	Z_{22}	$\dfrac{-Y_{21}}{\Delta Y}$	$\dfrac{Y_{11}}{\Delta Y}$	$\dfrac{-h_{21}}{h_{22}}$	$\dfrac{1}{h_{22}}$	$\dfrac{g_{21}}{g_{11}}$	$\dfrac{\Delta g}{g_{11}}$	$\dfrac{1}{C}$	$\dfrac{D}{C}$
Y	$\dfrac{Z_{22}}{\Delta Z}$	$\dfrac{-Z_{12}}{\Delta Z}$	Y_{11}	Y_{12}	$\dfrac{1}{h_{11}}$	$\dfrac{h_{12}}{h_{11}}$	$\dfrac{\Delta g}{g_{22}}$	$\dfrac{g_{12}}{g_{22}}$	$\dfrac{D}{B}$	$\dfrac{-\Delta T}{B}$
	$\dfrac{-Z_{21}}{\Delta Z}$	$\dfrac{Z_{11}}{\Delta Z}$	Y_{21}	Y_{22}	$\dfrac{h_{21}}{h_{11}}$	$\dfrac{\Delta h}{h_{11}}$	$\dfrac{-g_{21}}{g_{22}}$	$\dfrac{1}{g_{22}}$	$\dfrac{-1}{B}$	$\dfrac{A}{B}$
h	$\dfrac{\Delta Z}{Z_{22}}$	$\dfrac{Z_{12}}{Z_{22}}$	$\dfrac{1}{Y_{11}}$	$\dfrac{Y_{12}}{Y_{11}}$	h_{11}	h_{12}	$\dfrac{g_{22}}{\Delta g}$	$\dfrac{g_{12}}{\Delta g}$	$\dfrac{B}{D}$	$\dfrac{\Delta T}{D}$
	$\dfrac{-Z_{21}}{Z_{22}}$	$\dfrac{1}{Z_{22}}$	$\dfrac{Y_{22}}{Y_{11}}$	$\dfrac{\Delta Y}{Y_{11}}$	h_{21}	h_{22}	$\dfrac{-g_{21}}{\Delta g}$	$\dfrac{g_{11}}{\Delta g}$	$\dfrac{-1}{D}$	$\dfrac{C}{D}$
g	$\dfrac{1}{Z_{11}}$	$\dfrac{-Z_{12}}{Z_{11}}$	$\dfrac{\Delta Y}{Y_{22}}$	$\dfrac{Y_{12}}{Y_{22}}$	$\dfrac{h_{22}}{\Delta h}$	$\dfrac{-h_{12}}{\Delta h}$	g_{11}	g_{12}	$\dfrac{C}{A}$	$\dfrac{-\Delta T}{A}$
	$\dfrac{Z_{21}}{Z_{11}}$	$\dfrac{\Delta Z}{Z_{11}}$	$\dfrac{-Y_{21}}{Y_{22}}$	$\dfrac{1}{Y_{22}}$	$\dfrac{-h_{21}}{\Delta h}$	$\dfrac{h_{11}}{\Delta h}$	g_{21}	g_{22}	$\dfrac{1}{A}$	$\dfrac{B}{A}$
T	$\dfrac{Z_{11}}{Z_{21}}$	$\dfrac{\Delta Z}{Z_{21}}$	$\dfrac{-Y_{22}}{Y_{21}}$	$\dfrac{-1}{Y_{21}}$	$\dfrac{-\Delta h}{h_{21}}$	$\dfrac{-h_{11}}{h_{21}}$	$\dfrac{1}{g_{21}}$	$\dfrac{g_{22}}{g_{21}}$	A	B
	$\dfrac{1}{Z_{21}}$	$\dfrac{Z_{22}}{Z_{21}}$	$\dfrac{-\Delta Y}{Y_{21}}$	$\dfrac{-Y_{11}}{Y_{21}}$	$\dfrac{h_{22}}{h_{21}}$	$\dfrac{-1}{h_{21}}$	$\dfrac{g_{11}}{g_{21}}$	$\dfrac{\Delta Y}{g_{21}}$	C	D

$\Delta Z = Z_{11}Z_{22} - Z_{12}Z_{21}$

$\Delta Y = Y_{11}Y_{22} - Y_{12}Y_{21}$

$\Delta g = g_{11}g_{22} - g_{12}g_{21}$

$\Delta h = h_{11}h_{22} - h_{12}h_{21}$

$\Delta T = AD - BC$

18 Operational Amplifier Circuits

The output voltage for circuits with ideal op amps is listed in the following table.

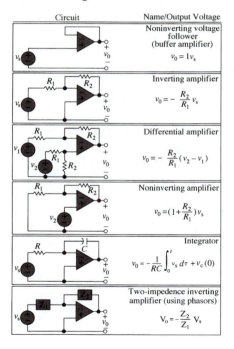

Circuit	Name/Output Voltage
	Noninverting voltage follower (buffer amplifier) $v_0 = 1v_s$
	Inverting amplifier $v_0 = -\dfrac{R_2}{R_1}v_s$
	Differential amplifier $v_0 = -\dfrac{R_2}{R_1}(v_2 - v_1)$
	Noninverting amplifier $v_0 = (1 + \dfrac{R_2}{R_1})v_s$
	Integrator $v_0 = -\dfrac{1}{RC}\displaystyle\int_0^t v_s\,d\tau + v_c(0)$
	Two-impedance inverting amplifier (using phasors) $V_o = -\dfrac{Z_2}{Z_1}V_s$

19 Electric Signals

An *electric signal* is a voltage or current varying with time in a manner that conveys information. A *signal* is defined as a real-valued function of time. By real valued we mean that for any fixed value of time, the value of the signal at that time is a real number.

Name	Equation	Waveform
1. Continuous or DC	$v(t) = V_0$	
2. Step	$v(t) = 0 \quad t < 0$ $\quad\quad\quad = V_0 \quad t \geq 0$	
3. Decaying Exponential	$v(t) = 0 \quad\quad t < 0$ $\quad\quad\quad = V_0 e^{-at} \quad t \geq 0$	
4. Ramp	$v(t) = 0 \quad t < 0$ $\quad\quad\quad = Kt \quad t \geq 0$	
5. Pulse	$v(t) = V_0 \quad 0 \leq t \leq t_1$ $\quad\quad\quad = 0 \quad\quad$ elsewhere	
6. Sinusoid	$v(t) = V_0 \sin(\omega t + \theta)$	

Note: Voltage is used to designate the waveform equation; current could be used equally as well.

20 Feedback Systems

A block diagram of a negative feedback system is shown in Figure 20.1. The overall system input–output transfer function $T(s)$ is

$$T(s) = \frac{C(s)}{R(s)} = \frac{G(s)}{1 + GH(s)}$$

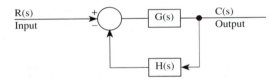

FIGURE 20.1. A negative feedback system.

21 Frequency Response

1. Bode Plots

The *Bode plot* is a chart of gain in decibels and phase in degrees versus the logarithm of frequency:

$$\text{logarithmic gain} = 20\log_{10} H$$

where $\mathbf{H} = H\underline{/\theta}$.

2. Resonant Circuits

A resonant circuit is a combination of frequency-sensitive elements to provide a frequency-selective response.

The quality factor for a parallel RLC resonant circuit is

$$Q = \omega_0 CR = \frac{R}{\omega_0 L}$$

and the resonant frequency is

$$\omega_0 = \frac{1}{\sqrt{LC}}$$

The *bandwidth* of a frequency-selective circuit is the frequency range between the points where the magnitude of the gain drops to $1/\sqrt{2}$ times the maximum value. Therefore,

$$B = \omega_2 - \omega_1$$

$$= \frac{\omega_0}{Q}$$

For a series RLC resonant circuit we have a quality factor

$$Q = \frac{\omega_0 L}{R} = \frac{1}{\omega_0 RC}$$

For $Q > 10$ and small deviations from ω_0, where the deviation is

$$\delta = \frac{\omega - \omega_0}{\omega_0}$$

we have the transfer function for the series or parallel resonant circuit:

$$\mathbf{H} = \frac{1}{1 + jQ(\omega/\omega_0 - \omega_0/\omega)}$$

$$= \frac{1}{1 + 2Q\delta}$$

22 System Response

1. The Convolution Theorem

The output of a circuit or a system with a transfer function $H(s)$ as shown in Figure 22.1 is

$$y(t) = \mathscr{L}^{-1}[H(s)R(s)]$$

$$= \int_0^t h(\tau)r(t-\tau)\,d\tau$$

2. The Impulse Function

An *impulse function* $\delta(t)$ is a pulse of infinite amplitude for an infinitesimal time whose area $\int_{-\infty}^{\infty}\delta(t)\,dt$ is finite. The impulse function is defined as

$$\delta(t) = 0 \quad \text{for } t \neq 0$$

and

$$\int_{-\infty}^{\infty} \delta(t)\,dt = 1$$

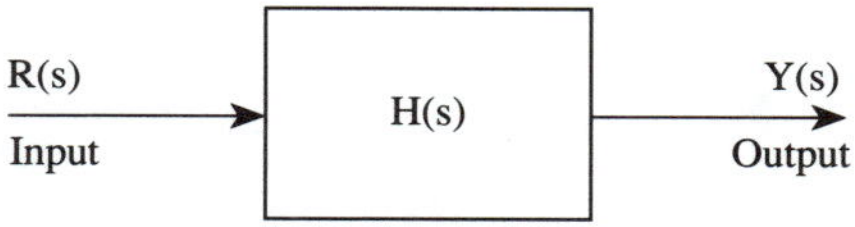

FIGURE 22.1.

3. Impulse Response

The impulse response is the output $y(t)$ of the system shown in Figure 22.1 with an input $r(t) = \delta(t)$. The Laplace transform of $r(t) = \delta(t)$ is

$$R(s) = 1$$

Then, the output $y(t)$, called the *impulse response*, is

$$y(t) = \mathscr{L}^{-1}[H(s)R(s)]$$
$$= \mathscr{L}^{-1}[H(s)]$$
$$= h(t)$$

4. Stability

A circuit or system is said to stable when the response to a bounded input signal is a bounded output signal. Thus, for a stable system $H(s)$ we require a finite or bounded impulse response:

$$\lim_{t \to \infty} |h(t)| = \text{finite}$$

For a linear circuit or system, we require for stability that all the poles of $H(s)$ be in the left-hand s-plane.

23 Fourier Series

A *Fourier series* is an accurate representation of a periodic signal which consists of the sum of sinusoids at the fundamental and harmonic frequencies.

The expression for a finite sum of harmonically related sinusoids called a *Fourier series* is

$$f(t) = a_0 + \sum_{n=1}^{N} a_n \cos n\omega_0 t$$

$$+ \sum_{n=1}^{N} b_n \sin n\omega_0 t$$

where $\omega_0 = 2\pi/T$ and a_0, a_n, and b_n (all real) are called the *Fourier trigonometric* coefficients.

An alternative form called the *exponential form* of the Fourier series is

$$f(t) = \sum_{-\infty}^{\infty} C_n e^{jn\omega_0 t}$$

where C_n are the complex (phasor) coefficients defined by

$$C_n = \frac{1}{T} \int_{t_0}^{t_0 + T} f(t) e^{-jn\omega_0 t}\, dt$$

where $C_n = C_{-n}^{*}$.

Because C_n are complex numbers, we may write

$$C_n = |C_n| \underline{/\theta_n}$$

and we may plot $|C_n|$ and $\underline{/\theta_n}$ as the amplitude spectrum and the phase spectrum.

24 Fourier Transform

The Fourier transform of $f(t)$ is

$$F(j\omega) = \int_{-\infty}^{\infty} f(t)e^{-j\omega t}\,dt$$

The inverse Fourier transform is

$$f(t) = \frac{1}{2\pi}\int_{-\infty}^{\infty} F(j\omega)e^{-j\omega t}\,d\omega$$

The spectrum of a signal $f(t)$ is its Fourier Transform $F(j\omega)$

25 Paresval's Theorem

The energy absorbed by a 1-Ω resistor with a voltage $v(t)$ across it is

$$w = \int_{-\infty}^{\infty} v^2(t)$$

$$= \frac{1}{2\pi} \int_{-\infty}^{\infty} |V(j\omega)|^2 \, d\omega$$

26 Static Electric Fields

1. Unit Vectors and Coordinate Systems

The unit vectors for the Cartesian (rectangular) system shown in Figure 26.1(a) are

$$\mathbf{a}_x, \quad \mathbf{a}_y, \quad \mathbf{a}_z$$

and all three vectors are constant.

The unit vectors for the cylindrical coordinate system shown in Figure 26.1(b) are

$$\mathbf{a}_\rho, \quad \mathbf{a}_\phi, \quad \mathbf{a}_z$$

where $\mathbf{a}_z$ is constant.

The unit vectors for the spherical coordinate system shown in Figure 26.1(c) are

$$\mathbf{a}_r, \quad \mathbf{a}_\theta, \quad \mathbf{a}_\phi$$

2. Coulomb's Law

For two-point charges Q_1, the source of the field force $\mathbf{F}$, and Q we have the force on Q as

$$\mathbf{F} = \frac{Q_1 Q}{4\pi\epsilon_0 R^2}\mathbf{a}_R \quad (\text{N})$$

where $\mathbf{a}_R$ is the vector of unit length pointing from Q_1

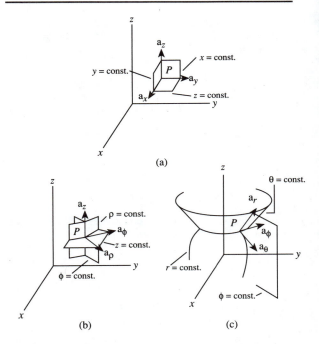

FIGURE 26.1. Unit vectors for (a) Cartesian, (b) cylindrical, and (c) spherical coordinates.

to Q, $\epsilon_0 = 10^{-9}(36\pi)$, and R is the distance between the charges.

Electric Field Intensity

The electrostatic field intensity is defined as the force on Q when $Q = 1$ C so

$$E = \frac{Q_1}{4\pi\epsilon_0 R^2}\mathbf{a}_R \quad (\text{V/m})$$

and

$$\mathbf{F} = Q\mathbf{E} \quad (\text{N})$$

3. Gauss' Law

Electric flux density, **D**, is

$$\mathbf{D} = \epsilon_0 \mathbf{E} \quad (\text{C/m}^2)$$

Gauss' law states that the net flux of **D**, or electric flux ψ passing through a surface is equal to the net positive charge enclosed within the surface and thus

$$\psi = \oint \mathbf{D}_s \cdot d\mathbf{S} = Q$$

where $\mathbf{D}_s$ is the value of **D** at the surface and $d\mathbf{S}$ is the surface element.

4. Maxwell's Equation (Electrostatics)

The electric flux per unit volume leaving a vanishingly small volume unit is equal to the volume charge density there:

$$\text{div}\,\mathbf{D} = \rho$$

where div is divergence and ρ is a volume charge density. Using $\text{div}\,\mathbf{D} = \mathbf{\nabla}\cdot\mathbf{D}$, we have

$$\mathbf{\nabla}\cdot\mathbf{D} = \rho$$

5. Poisson's Equation

$$\nabla \cdot \nabla V = -\frac{\rho}{\epsilon}$$

or

$$\nabla^2 V = -\frac{\rho}{\varepsilon}$$

where $\mathbf{E} = -\nabla V$.

6. Current Density

The current density $\mathbf{J}$ is related to the electric field $\mathbf{E}$ for a metallic conductor as

$$\mathbf{J} = \sigma \mathbf{E}$$

where σ is the conductivity of the conductor.

The current density $\mathbf{J}$ is a convection current

$$\mathbf{J} = \rho \mathbf{v}$$

where $\mathbf{v}$ is a velocity vector and ρ is the volume charge density.

27 Static Magnetic Fields

1. Biot–Savart Law

A current I flowing in a differential vector length $d\mathbf{L}$ results in a magnetic field intensity $\mathbf{H}$ as

$$d\mathbf{H} = \frac{I\,d\mathbf{L} \times \mathbf{a}_R}{4\pi R^2} \quad \text{(A/m)}$$

Expressed in terms of current density $\mathbf{J}$, we have

$$\mathbf{H} = \int_{\text{volume}} \frac{\mathbf{J} \times \mathbf{a}_R}{4\pi R^2}\,dv$$

2. Ampere's Law

The line integral of $\mathbf{H}$ about any closed path is equal to the direct current enclosed by that path:

$$\oint \mathbf{H} \cdot d\mathbf{L} = I$$

3. Maxwell's Equations for Static Fields

$$\boldsymbol{\nabla} \times \mathbf{H} = \mathbf{J}$$

and

$$\boldsymbol{\nabla} \times \mathbf{E} = 0$$

4. Stokes' Theorem

$$\oint \mathbf{H} \cdot d\mathbf{L} = \int_{\text{surface } S} (\mathbf{\nabla} \times \mathbf{H}) \cdot d\mathbf{S}$$

5. Magnetic Flux Density

Magnetic flux density $\mathbf{B}$ in free space is

$$\mathbf{B} = \mu_0 \mathbf{H} \quad \text{(T)}$$

where T is teslas and $\mu_0 = 4\pi \times 10^{-7}$ H/m.

Then, the divergence theorem provides

$$\mathbf{\nabla} \cdot \mathbf{B} = 0$$

28 Maxwell's Equations

1. Maxwell's Equations for Static Fields

Differential Form	Integral Form
$\nabla \cdot \mathbf{D} = \rho$	$\oint_{\text{surface}} \mathbf{D} \cdot d\mathbf{S} = Q = \int_{\text{volume}} \rho \, dv$
$\nabla \times \mathbf{E} = 0$	$\oint \mathbf{E} \cdot d\mathbf{L} = 0$
$\nabla \times \mathbf{H} = \mathbf{J}$	$\oint \mathbf{H} \cdot d\mathbf{L} = I = \int_{\text{surface}} \mathbf{J} \cdot d\mathbf{S}$
$\nabla \cdot \mathbf{B} = 0$	$\oint \mathbf{B} \cdot d\mathbf{S} = 0$

2. Maxwell's Equations for Time-Varying Fields

$$\nabla \times \mathbf{E} = -\frac{\partial \mathbf{B}}{\partial t}$$

$$\nabla \times \mathbf{H} = \mathbf{J} + \frac{\partial \mathbf{D}}{\partial t}$$

$$\nabla \cdot \mathbf{D} = \rho$$

$$\nabla \cdot \mathbf{B} = 0$$

29 Semiconductors

1. Current in a Semiconductor

In semiconductors both holes and electrons contribute to electrical conduction. With an applied electric field **E**, the expression for current density is

$$J = (n\mu_n + p\mu_p)e\mathbf{E} = \sigma\mathbf{E}$$

where n and p are the concentrations of electrons and holes (number/m^3) and μ_n and μ_p are the corresponding mobilities. Conductivity depends on the number of charge carriers and their mobility; for a semiconductor, the conductivity is

$$\sigma = (n\mu_n + p\mu_p)e$$

In a pure semiconductor the number of holes is just equal to the number of conduction electrons, or

$$n = p = n_i$$

where n_i is the *intrinsic* concentration.

In a doped semiconductor,

$$np = n_i^2$$

In words, the product of electron and hole concentrations is a constant; if one is increased (by doping), the other must decrease. If the doping concentration is nonuniform, the concentration of charged particles is also nonuniform, and it is possible to have charge motion by the mechanism called *diffusion*. The *diffusion current* is proportional to the concentration gradient dn/dx. The diffusion current density due to electrons is given by

$$J_n = eD_n \frac{dn}{dx}$$

where D_n is the diffusion constant for electrons (m^2/s).

The diffusion current density due to nonuniform concentrations of randomly moving electrons and holes is

$$J = J_n + J_p = eD_n \frac{dn}{dx} - eD_p \frac{dp}{dx}$$

2. *Semiconductor Diodes*

A semiconductor diode conducts forward current with a small forward voltage drop across the device, simulating a closed switch. The relationship between the forward current and forward voltage is a good approximation given by the Shockley diode equation.

$$i = I_s [e^x - 1]$$

where

$$x = \frac{eV}{kT}$$

and where I_s is the leakage current through the diode, e is electronic charge, k is Boltzman's constant, T is the temperature of the diode, and V is the voltage across the diode.

3. Field Effect Transistors

In a *junction field effect transistor* (JFET), the width of the depletion layers controls the conductance. For the JFET, the drain current in the constant-current region is

$$i_{DS} - I_{DSS}(1 - v_{GS}/V_p)^2$$

where i_{DS} is the drain current in the constant-current region, I_{DSS} is the value of i_{DS} with gate shorted to source, and V_p is the pinch-off voltage.

For an enhancement MOSFET, the transfer characteristic is

$$i_{DS} = K(v_{GS} - V_T)^2$$

where K is a device parameter and V_T is the turn-on or threshold voltage.

4. Bipolar Junction Transistors (BJT)

A bipolar junction transistor consists of two *pn* junctions in close proximity; normally, the emitter junction is forward biased, the collector reverse biased. In common-base operation, the collector current i_C is

$$i_c = -\alpha i_E + I_{CBO} \quad \text{where } \alpha \cong 1$$

where I_{CBO} is the collector cutoff current and α is the forward current-transfer ratio. In common-emitter operation, a small base current controls the relatively larger collector current to achieve current amplification:

$$i_C = \beta i_B + I_{\text{CEO}} \quad \text{where } \beta = \frac{\alpha}{1 - \alpha}$$

where i_B is the base current and I_{CEO} is the collector cutoff current in the common-emitter configuration.

30 Digital Logic

1. AND *Gate*

A *logic gate* is a device that controls the flow of information, usually in the form of pulses. The symbol for an AND gate is shown in Figure 30-1. A·B is read "A AND B." As indicated in the truth table, an output appears only when there are inputs at A AND B.

2. OR *Gate*

The symbol for an OR gate is shown in Figure 30.2, where A + B is read "A OR B." As indicated in the truth table, the output is 1 if input A OR input B is 1. For no input, the output is zero (0).

3. NOT *Gate*

The logic NOT is represented by the symbol in Figure 30.3, where $\overline{A}$ is read "NOT A." As indicated in the

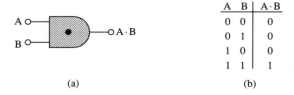

A	B	A·B
0	0	0
0	1	0
1	0	0
1	1	1

(a)   (b)

FIGURE 30.1. (a) Symbol and (b) truth table for the AND gate.

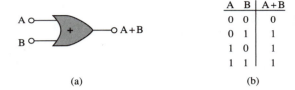

A	B	A+B
0	0	0
0	1	1
1	0	1
1	1	1

(a)   (b)

FIGURE 30.2. A two-input OR gate (a) symbol and (b) truth table.

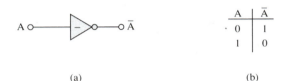

A	$\overline{A}$
0	1
1	0

(a)   (b)

FIGURE 30.3. A NOT gate (a) symbol and (b) truth table.

truth table, the NOT element is an *inverter*; the output is the *complement* of the single input.

4. NAND *Gate*

The NAND gate is defined by the truth table of Figure 30.4. The circle on the NAND element symbol and the bar on the $\overline{A \cdot B}$ output indicate the inversion process

5. Exclusive-OR *Gate*

The Exclusive-OR operation is $(A + B)\overline{AB}$ as shown in Figure 30.5. The Exclusive-OR gate is used so fre-

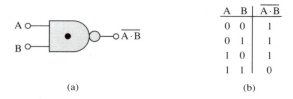

A	B	$\overline{A \cdot B}$
0	0	1
0	1	1
1	0	1
1	1	0

(a)   (b)

FIGURE 30.4. The NAND gate (a) symbol and (b) truth table.

A	B	$A \oplus B$
0	0	0
0	1	1
1	0	1
1	1	0

(a)   (b)

FIGURE 30.5. The Exclusive-OR gate (a) symbol and (b) truth table.

quently that it is represented by the special symbol $\oplus$ defined by

$$A \text{ XOR } B = A \oplus B = (A + B)\overline{AB}$$

6. *DeMorgan's Theorem*

DeMorgans Theorem states that "to obtain the inverse of any Boolean function, invert all variables and replace all ORs by ANDs and all ANDs by ORs."

The first DeMorgan theorem says that a NOR gate $(\overline{A+B})$ is equivalent to an AND gate with NOT circuits in the inputs $(\overline{A} \cdot \overline{B})$. The second says that a NAND gate $(\overline{A \cdot B})$ is equivalent to an OR gate with NOT circuits in the inputs $(\overline{A} + \overline{B})$.

31 Communication Systems

1. Half-Power Bandwidth

The constancy of the magnitude $|H(j\omega)|$ of a system is specified by a parameter called its bandwidth, B, and is defined as the interval of positive frequencies over which $|H(\omega)|$ remains within 3 dB (with $1/\sqrt{2}$ in voltage or $\frac{1}{2}$ in power).

2. The Sampling Theorem

A real-valued band-limited signal having no spectral components above a frequency B (H_z) is determined uniquely by its values at uniform intervals spaced no greater than $1/2B$ seconds apart.

For a signal $x(t)$ with a Fourier Transform $X(f)$, where $X(f)$ is assumed zero for $f \geq B$, the signal is recoverable from a sampling frequency f_s:

$$f_s \geq 2B$$

3. Amplitude Modulation

The equation of a general sinusoidal (carrier) signal can be written as

$$y(t) = a(t)\cos[\omega_c t + \phi(t)]$$

where we assume $a(t)$ and $\phi(t)$ vary slowly compared to $\omega_c t$. The term $a(t)$ is called the envelope of the signal, ω_c is the carrier frequency, and $\phi(t)$ is the phase modulation of $y(t)$.

The modulating signal $x(t)$ provides an amplitude-modulated carrier signal as

$$y(t) = kx(t) \cos \omega_c t$$

where $\phi(t) = 0$ and k is a constant.

4. Phase and Frequency Modulation

The modulating signal $x(t)$ can be used to modulate the frequency or phase of the carrier signal as

$$y(t) = A \cos \theta(t)$$

where A is a constant.

The relation between the instantaneous angular rate $\omega(t)$ and $\phi(t)$ is

$$\theta(t) = \int_0^t \omega(\tau)\, d\tau + \theta_0$$

or

$$\omega(t) = \frac{d\theta}{dt}$$

Phase modulation is obtained when

$$\theta(t) = \omega_c t + k_1 x(t) + \theta_0$$

and $x(t)$ is the modulating signal.

191

Frequency modulation is obtained when

$$\omega(t) = \omega_c + k_2 x(t)$$

5. A Measure of Information

The measure of information associated with an event A occurring with probability P_A is

$$I_A = \log \frac{1}{P_A}$$

with $\log_2$ (base 2).

6. Average Information (Entropy)

The average information, called the entropy H, of a message is

$$H = \sum_{i=1}^{n} P_i \log_2 \frac{1}{P_i} \qquad \text{bits}$$

7. Channel Capacity (Shannon's Theorem)

The limiting rate of information transmission through a channel is called the channel capacity. For a source with an available alphabet of α discrete messages, the maximum entropy of the source is $\log_2 \alpha$ bits, and if T is the transmission time of each message, the channel capacity C is

$$C = \frac{1}{T} \log_2 \alpha \qquad \text{bits/second}$$

when the messages are equally probable and statistically independent.

Index